T0261581

Making and Breaking Mathematical Sense

Making and Breaking Mathematical Sense

HISTORIES AND PHILOSOPHIES OF MATHEMATICAL PRACTICE

■ ■ ■ ■ ■ ■ ■

Roi Wagner

PRINCETON UNIVERSITY PRESS

PRINCETON AND OXFORD

Published by Princeton University Press, 41 William Street, Princeton, New Jersey 08540
In the United Kingdom: Princeton University Press, 6 Oxford Street, Woodstock, Oxfordshire
OX20 1TR

press.princeton.edu
Library of Congress Cataloging-in-Publication Data

Names: Wagner, Roi, 1973–
Title: Making and breaking mathematical sense : histories and philosophies of mathematical
    practice / Roi Wagner.
Description: Princeton : Princeton University Press, [2017] | Includes bibliographical
    references and index.
Identifiers: LCCN 2016022844 | ISBN 9780691171715 (hardcover : alk. paper)
Subjects: LCSH: Mathematics—Philosophy—History. | Mathematics—History
Classification: LCC QA8.4 .W334 2017 | DDC 510.1—dc23 LC record available at
    https://lccn.loc.gov/2016022844

British Library Cataloging-in-Publication Data is available

This book has been composed in Linux Libertine O and Myriad Pro

Printed on acid-free paper. ∞

Printed in the United States of America

10 9 8 7 6 5 4 3 2 1

To Rhone, who's too good for some cliché dedication

Contents

■ ■ ■ ■ ■ ■ ■ ■

Acknowledgments

■ ■ ■ ■ ■ ■ ■ ■ ■ ■ ■ ■ ■ ■

THIS BOOK INTEGRATES THE FRUITS of research and thinking that have spanned some ten years of work. Many people contributed to various parts of this research, either before or after it became a book project, with comments or encouragement. I would like to thank them wholeheartedly and alphabetically: Yemima Ben-Menachem, Anat Biletzki, Arianna Borrelli, Lin Chalozin-Dovrat, Leo Corry, Paul Ernest, Menachem Fisch, Gideon Freudenthal, Rachel Giora, Emily Grosholz, Eitan Grossman, Sandra Harding, Albrecht Heeffer, Jens Høyrup, Victor Katz, Stav Kaufman Raviv, Jay Lemke, Reviel Netz, Adi Ophir, Arkady Plotnitsky, Yossef Schwartz, Sabetai Unguru, Warren Van Egmond. If I have failed to mention anyone here it is a failure of memory and not a lack of gratitude.

I can't leave out the people who taught me mathematics: Vitali Milman and Efim Gluskin. Even if they end up agreeing with nothing in this book, my understanding of how mathematics works would be even poorer if it weren't for them.

Finally, I thank my editor, Vickie Kearn, for her enthusiastic support.

Making and Breaking Mathematical Sense
■ ■ ■ ■ ■ ■ ■

Introduction
∎∎∎∎∎∎∎∎∎∎

What Philosophy of Mathematics Is Today

One of the starting points for this book project was a job talk that I presented at the philosophy department of some research university. I discussed how mathematical signs shift their meanings and described mathematical processes of sense making, some of which are covered in this book. One of the department professors commented succinctly that "this is not philosophy of mathematics." He explained that a philosopher of mathematics should take the notions and terms that we or our historical sources use when discussing mathematics and provide them with some sort of rational reconstruction. That was not what I was doing.

In a sense, this professor was right. What I did was not philosophy of mathematics as usually practiced today. This can be easily verified. I went through the top 150 entries in the *Philosopher's Index* with the words "mathematics" or "mathematical" in their abstracts that were posted over the last couple of years. This showed that the most popular debate in contemporary philosophy of mathematics, representing almost 40 percent of research production, is how to describe the mode of existence of mathematical entities, especially in the context of their application to natural sciences (the *PhilPapers* bibliographies suggest a similar proportion).

So the main problem that bothers mainstream philosophy of mathematics today has to do with the kind of reality attributed to mathematical objects and statements. Indeed, in most situations, one speaks of mathematical objects and statements like one speaks of other scientific objects and statements. But there are obvious problems with this way of speaking, because mathematical objects are hard to tie down to spatio-temporal phenomena, and the claims that involve mathematical objects are therefore hard to conceive as truly referential, as realists require. At the same time, many mathematical claims are highly applicable to empirical phenomena, which makes it difficult to think of them as contingent constructs of reason, as nominalists tend to do.

So the main contemporary philosophical task in mainstream philosophy of mathematics is to trace a rational account of the terms "true" and "exist" so as to allow their consistent use in both science and mathematics, and, at the same time, respect the common usage of these terms and common mathematical habits. But this latter pair of constraints is in conflict: we may either redefine "true" and "exist" creatively to generate consistency over their use in science and mathematics, but lose contact with common usage, or we can hold on to common usage, but face obstacles when applying these terms to real, existing mathematical practice. The philosophical debate is thus about stretching the terms "true" and "exist" in ways that cover the most important aspects of common usage and conceptual consistency. Since deciding what is "most important" involves nonconsensual prioritizing, the debate continues to spin.

This analytic approach refers to an established canon of philosophies of mathematics. The main references are the following: Plato's transcendent and ideal mathematical forms recollected through empirical experience and dialectical reason; Kant's view of mathematics as a science of the forms *through* which we organize time and space (a middle ground between empiricist accounts of mathematics in terms of observations *in* time and space and rationalist accounts in terms of pure nonempirical reason); the logicist attempt to reduce mathematics to logic as the science of pure reason; the intuitionist attempt to confine mathematics to what can be actually constructed in our minds (or, more exactly, in a Kantian-like form of temporal intuition); the formalist articulation of mathematics as a system of meaningless signs subject to purely syntactic rules whose consistency is to be analyzed by means of a constructive and finitary logic; the logical positivist articulation of mathematics as a system of syntactic logical truths used to tie together empirical observations; and the empiricist-holist view of mathematics and empirical science as an inseparable continuum. This canon provides the backdrop for the contemporary search for a satisfactory articulation of mathematical objectivity and truth.

This entire canon revolves around a foundational question: "To what kind of ontological ground can we *reduce* mathematics?" It is tied together by a search for a unified ontological substrate and for a unified language to discuss it. So from the point of view of this tradition, the professor who criticized my presentation was right. I was not doing philosophy of mathematics, and I continue not to do it in this book. (Obviously, I didn't get the job.)

What Else Philosophy of Mathematics Can Be

Before trying to articulate a different kind of philosophy of mathematics, I should explain who this philosophy is meant for. As I see it, there are three main target groups for the philosophy of mathematics: philosophers, mathematicians, and people who engage with mathematics less intensively in their professional and daily lives.

Philosophers are usually interested in mathematics as a test case for some general philosophical system. Since it is not just any test case, but one that has some extreme characteristics (for instance, it is seen as an extremely rigorous branch of knowledge), mathematics is considered a very important test case. Since the interest in mathematics is usually entangled with science and logic, and since these domains are favorites of the analytical tradition, philosophers' philosophy of mathematics is usually analytic—hence the current focus on analytic versions of realism versus nominalism and questions concerning math-science relations.

Mathematicians, as far as I can see, are not terribly interested in the philosophy of mathematics. They often have philosophical views, but they are usually not very keen on challenging or developing them—they don't usually consider this as worthy of too much effort. They're also very suspicious of philosophers. Indeed, mathematicians know better than anyone else what it is that they're doing. The idea of having a philosopher lecture them about it feels kind of silly, or even intrusive.

So we turn to people who have something to do with mathematics in their professional or daily lives, but are not focused on mathematics. Such people often have some sort of vague, sometimes naïve, conceptions of mathematics. One of the most striking manifestations of these folk views is the following: If I say something philosophical that people don't understand, the default assumption is that I use big pretentious words to cover small ideas. If I say something mathematical that people don't understand, the default assumption is that I'm saying something so smart and deep that they just can't get it.

There's an overwhelming respect for mathematics in academia and in wider circles. So much so that bad, trivial, and pointless forms of mathematization are often mistaken for important achievements in the social sciences, and sometimes in the humanities as well. It is often assumed that all ambiguities in our vague verbal communication disappear once we switch to mathematics, which is supposed to be purely univocal and absolutely true. But a mirror image of this approach is also common. According to this view, mathematics is a purely mechanical, inhuman, and irrelevantly abstract form of knowledge.

I believe that the philosophy of mathematics should try to confront such naïve views. To do that, one doesn't need to reconstruct a rational scheme underlying the way we speak of mathematics, but rather paint a richer picture of mathematics, which tries to affirm, rather than dispel, its ambiguities, humanity, and historicity.

This approach represents a minority strand in the philosophy of mathematics—but a growing minority. There's a whole tradition that's been coming together since the 1980s (if not earlier), which, today, is referred to under the title "philosophy of mathematical practice" (Azzouni 1994; Ernest 1998; Hersh 1997; Van Kerkhove 2009; Van Kerkhove and Van Bendegem 2007; Mancosu 1996, 2011; Rotman 2000; Tymoczko 1998). This tradition tries to explain what it is that mathematicians do when they do mathematics, and to shift the focus from "what it is" to "how it works." This shift does not quite exclude the question of "what mathematics is"; rather, it asks "what it is in motion," as it is being produced, understood, interpreted, and applied, rather than "what it is at rest," when we try look at it as a complete, given object or stratum of reality/discourse.

When proponents of this school of philosophy of mathematics look at some historical or contemporary practices that are problematic according to contemporary formal standards (for example, the use of infinitesimals or argument by diagrams), they do not try to reconstruct them in more rigorous terms; rather they look at whatever it is that mathematicians do, even when what they do is not formally rigorous (I, personally, tend to focus on semiosis—how mathematical signs accumulate and change meanings).

This latter branch of the philosophy of mathematics is highly descriptive and deeply entangled with the history and sociology of mathematics. Indeed, when we look at what mathematicians do, we find that their practice changes historically and is irreducibly embedded in social institutions. Due to the descriptive and concrete flavor of this kind of work, many see it as unworthy of the title "philosophical." It often has to call itself "science studies" to survive outside the traditional disciplinary boundaries of philosophy, history, and sociology. In many ways, what I do with mathematical texts is more similar to what some researchers in literature departments do with literary texts than to what people in philosophy departments do (this obviously has to do with the fact that some branches of continental philosophy were exiled to literature departments).

Nevertheless, I believe that this form of dealing with mathematics, regardless of whether we choose to call it philosophy or not, is genuinely important today. The approach that I'm promoting here can help a general academic readership reform some folk views of mathematics, and reposition mathematics as a humanely accessible endeavor, enjoying many unique characteristics, but still comparable to other branches of knowledge.

The approach advocated here is gaining more and more interest from philosophers, and is less alienating to mathematicians than mainstream philosophical accounts. Hopefully, it can generate a discourse that will draw together philosophers, mathematicians, and nonspecialists, so as to reintegrate the scattered sectarian debates on the present and future of mathematics into a lively and more pertinent cross-cultural conversation.

The uncritical idolizing of mathematics as *the best* model of knowledge, just like the opposite trend of disparaging mathematics as mindless drudgery, are both detrimental to the organization and evaluation of contemporary academic knowledge. Instead, mathematics should be appreciated and judged as one among many practices of shaping knowledge. Understanding what this practice consists of would allow the academic community to give mathematics its due credit and place within and outside the academic system.

But before we go any further, in order to give a more concrete sense of what I am trying to deal with, let's consider the following vignette.

A Vignette: Option Pricing and the Black-Scholes Formula

The point of the following vignette is to give a concrete example of how mathematics relates to its wider scientific and practical context. It will show that mathematics has force, and that its force applies even when actual mathematical claims do not quite work as descriptions of reality. The rest of this book will then try to philosophize about this force: where it comes from, how it works, and how it interacts with other forces.

The context of this vignette is option pricing. An "option" is the right (but not the obligation) to make a certain transaction at a certain cost at a certain time. For example, I could own the option to buy 100 British pounds for 150 US dollars three months from today. If I own this option, and three months from today 100 pounds are worth more than 150 dollars, I'd be likely to use the option. If 100 pounds turn out to be worth less than 150 dollars, I will most probably simply discard it.

Such options could be used as insurance. The preceding option, for example, would insure me against a drop in the dollar-pound exchange rate, if I needed such insurance. It could also serve as a simple bet for financial gamblers. But what price should one put on this kind of insurance or bet?

There are two narratives to answer this question. The first says that until 1973, no one really knew how to price such options, and prices were determined by supply, demand, and guesswork. More precisely, there existed some reasoned means to price options, but they all involved putting a price on the risk one was willing to take, which is a rather subjective issue.

In two papers published in 1973, Fischer Black and Myron Scholes, followed by Robert Merton, came up with a reasoned formula for pricing options that did not require putting a price on risk. This feat was deemed so important that in 1997 Scholes and Merton were awarded the Nobel Prize in economics for their formula (Black had died two years earlier). Indeed, "Black, Merton and Scholes thus laid the foundation for the rapid growth of markets for derivatives in the last ten years"—at least according to the Royal Swedish Academy press release (1997).

But there's another way to tell the story. This other way claims that options go back as far as antiquity, and option pricing has been studied as early as the seventeenth century. Option pricing formulas were established well before Black and Scholes, and so were various means to factor out putting a price on risk (based on something called put-call parity rather than the Nobel-winning method of dynamic hedging, but we can't go into details here). Moreover, according to this narrative, the Black-Scholes formula simply doesn't work and isn't used (Derman and Taleb 2005; Haug and Taleb 2011).

If we wanted to strike a compromise between the two narratives, we could say that the Black-Scholes model was a new and original addition to existing models, and that it works under suitable ideal conditions, which are not always approximated by reality. But let's try to be more specific.

The idea behind the Black-Scholes model is to reconstruct the option by a dynamic process of buying and selling the underlying assets (in our preceding example, pounds and dollars). It provides an initial cost and a recipe that tells you how to continuously buy and sell these dollars and pounds as their exchange rate fluctuates over time in order to guarantee that by the time of the transaction, the money one has accumulated together with the 150 dollars dictated by the option would be enough to buy 100 pounds. This recipe depends on some clever, deep, and elegant mathematics.

This recipe is also risk free and will necessarily work, provided some conditions hold. These conditions include, among others, the capacity to always instantaneously buy and sell as many pounds/dollars as I want and a specific probabilistic model for the behavior of the exchange rate (Brownian motion with a fixed and known future volatility, where volatility is a measure of the fluctuations of the exchange rate).

The preceding two conditions do not hold in reality. First, buying and selling is never really unlimited and instantaneous. Second, exchange rates do not adhere precisely to the specific probabilistic model.

But if we can buy and sell fast enough, and the Brownian model is a good enough approximation, the pricing formula should work well enough. Unfortunately, prices sometimes follow other probabilistic models (with some infinite moments), where the Black and Scholes formula may fail to be even approximately true. The latter flaw is sometimes cited as an explanation for some of the recent market crashes—but this is a highly debated interpretation.

Another problem is that the future volatility (a measure of cost fluctuations from now until the option expires) of whatever the option buys and sells has to be known for the model to work. One could rely on past volatility, but when comparing actual option prices and the Black-Scholes formula, this doesn't quite work. The volatility rate that is required to fit the Black-Scholes formula to actual market option pricing is not simply the past volatility.

In fact, if one compares actual option prices to the Black-Scholes formula, and tries to calculate the volatility that would make them fit, it turns out that there's no single volatility for a given commodity at a given time. The cost of wilder options (for selling or buying at a price far removed from the present price) reflects higher volatility than the more tame options. So something is clearly empirically wrong with the Black-Scholes model, which assumes a fixed (rather than stochastic) future volatility for whatever the option deals with, regardless of the terms of the option.

So the Black-Scholes formula is nice in theory, but needn't work in practice. Haug and Taleb (2011) even argue that practitioners simply don't use it, and have simpler practical alternatives. They go as far as to say that the Black-Scholes formula is like "scientists lecturing birds on how to fly, and taking credit for their subsequent performance—except that here it would be lecturing them the wrong way" (101, n. 13). So why did the formula deserve a Nobel prize?

Looking at some informal exchanges between practitioners, one can find some interesting answers. The discussion I quote from the online forum *Quora* was headed by the question "Is the Black-Scholes Formula Just Plain Wrong?" (2014). All practitioners agree that the formula is not used as such. Many of them don't quite see it as an approximation either. But this does not mean that they think it is useless. One practitioner (John Hwang) writes:

Where Black-Scholes really shines, however, is as a common language between options traders. It's the oldest, simplest, and the most intuitive option pricing model around. Every option trader understands it, and it is easy to calculate, so it makes sense to communicate implied volatility [the volatility that would make the formula fit the actual price] in terms of Black-Scholes.... As a proof, the exchanges disseminate [Black-Scholes] implied volatility in addition to price data.

Another practitioner (Rohit Gupta) adds that this "is done because traders have better intuition in terms of volatilities instead of quoting various prices." In the same vein, yet another practitioner (Joseph Wang) added:

One other way of looking at this is that Black-Scholes provides something of a baseline that lets you compare the real world to a nonexistent ideal world.... Since we don't live in an ideal world, the numbers are different, but the Black-Scholes framework tells us *how different* the real world is from the idealized world.

So the model earned its renown by providing a common language that practitioners understand well, and allowing them to understand actual contingent circumstances in relation to a sturdy ideal.

Now recall that practitioners extrapolate the implied volatility by comparing the Black-Scholes formula to actual prices, rather than plug a given volatility into the formula to get a price. This may sound like data fitting. Indeed, one practitioner (Ron Ginn) states that "if the common denominator of the crowd's opinion is more or less Black-Scholes ... smells like a self fulfilling prophecy could materialize," or, put in a more elaborate manner (Luca Parlamento):

I just want to add that CBOE [Chicago Board Options Exchange] in early '70 was looking to market a new product: something called "options." Their issue was that how you can market something that no one can evaluate? You can't! You need a model that helps people exchange stuff, turn[s] out that the BS formula ... did the job. You have a way to make people easily agree on prices, create a liquid market and ... "why not" generate commissions.

The tone here is more sinister: the formula is useful because it's there, because it's a reference point that allows a market to grow around it.

But why did *this* specific formula attract the market, and become a common reference point, possibly even a self-fulfilling prophecy? Why not any of the other older or contemporary pricing practices, which are no worse? Why was this specific pricing model deemed Nobel worthy?

The answer, I believe, lies in the mathematics. The formula depends on a sound and elegant argument. The mathematics it uses is sophisticated, and enjoys a record of good service in physics, which imparts a halo of scientific prestige. Moreover, it is expressed in the language of an expressive mathematical domain that makes sense to practitioners (and, of course, it also came at the right time).

This is the force of mathematics. It's a language that the practitioners of the relevant niches understand and value. It feels well founded and at least ideally true. If it is sophisticated and comes with a good track record in other scientific contexts, it is assumed to be deep and somehow true. All this helps build rich practical networks around mathematical ideas, even when these ideas do not reflect empirical reality very well.

This book is about the force of mathematics, its origins and its unfolding. We all have some basic ideas about how and why mathematics works, and to what extent it is true or useful. But if we want to understand the surprising force of mathematics demonstrated in this vignette, we need to engage in a more careful analysis of mathematical practice.

Outline of This Book

The purpose of this book is to investigate what this force of mathematics builds on, and how it works in practice. To do that, I will discuss mathematics not only from the point of view of applications but also from the point of view of its production.

Chapter 1 introduces some histories of canonized philosophies of mathematics. These historical narratives are structured so as to highlight *not* the specific philosophical questions of the canon, but some overarching concerns that philosophical debates reflect. This serves to creatively rearrange the canon of the philosophy of mathematics and introduce some of the problems that this book will engage.

This articulation of overarching problems is reflected in chapter 2 by a real historical case study. This chapter attempts to flesh out the preceding chapter's problems by describing economical-mathematical practice with algebraic signs and subtracted numbers in the *abbaco* tradition of the Italian late Middle Ages and Renaissance. This chapter follows the vein of Wagner (2010b, 2010c), but it is thoroughly reorganized, and includes new material.

This leads us to chapter 3: a general outline of a philosophy of mathematical practice that forms the theoretical core of this book (a philosophically inclined reader who wants to read just one chapter of this book should probably go directly there). This chapter reflects on the function of mathematical statements (following Wittgenstein), their epistemological position, mathematical consensus, and mathematical interpretation and semiosis. The various positions expressed in the philosophical survey of the first chapter and the historical case study of the second are rearticulated as real constraints that apply to mathematical practice. Different mathematical cultures negotiate these constraints in different ways, and no single constraint serves as a final foundation. The chapter then proceeds to engage more mainstream notions of reality and truth of mathematical entities and statements (following Grosholz and Maddy), and suggests how a takeoff on Putnam's notion of relevance might relativize them. The fourth section of this chapter includes material revised from Wagner (2010a).

Chapter 4 attempts to reflect some of the ideas of the previous chapter with concrete case studies, focusing on problems of mathematical semiosis: how mathematical signs obtain and change their senses. The case studies are abridged and simplified versions of Wagner (2009b, 2009c), dealing with generating functions and the stable marriage problem in combinatorics. This opens up questions of how meaning is transferred within and across mathematical contexts—a question that belongs to the study of mathematical cognition.

Chapter 5 will articulate the preceding cognitive concerns in a more systematic manner. In order to introduce the notion of embodied mathematical cognition, I will first review the neuro-cognitive debate on the mental representation of numbers (focusing on Dehaene and Walsh). I will then present the cognitive theory of mathematical metaphor and suggest a rearticulation, based on Wagner (2013). This theory will be further enriched by an engagement with Walter Freeman's theory of meaning. We will conclude with an appropriation of Deleuze's *Logic of Sensation* to the context of mathematical practice, drawing on material from Wagner (2009a).

Chapter 6 will flesh out the cognitive problematic from the previous chapter with case studies of medieval and early modern geometric algebra adapted from Wagner (2013) and of the history of notions of infinity adapted from Wagner (2012). These case studies will demonstrate the limitations of the cognitive theory of mathematical metaphor in accounting for the formation of actual historical mathematical life worlds.

Chapter 7 will complement the discussion by thinking of mathematics not only as subject to constraints but also as feeding back into the reality that shapes it. A brief narrative will follow Fichte, Schelling, and Hermann Cohen to derive inspiration for rethinking the reality of ideas, and suggest how mathematics reforms the world where it lives. These philosophical approaches will lead us to offer a solution to Mark Steiner's formulation of Wigner's problem of the "unreasonable" applicability of mathematics to the natural sciences (or at least a reduction of the problem to a more containable intra-mathematical setting), elaborating an argument briefly outlined in Wagner (2012).

The book should be accessible to readers with a general interest in philosophy and mathematics. Some of the case studies and examples may require the equivalent of basic undergraduate calculus or linear algebra courses. Only a few scattered examples require higher mathematical training, and they can be skipped. The modular structure of the book should help readers avoid sections that are too theoretical or too technical for their taste.

I include various historic and contemporary case studies, some of which may appear rather strange to contemporary readers. My purpose is to recall that mathematics was and can be different from the mathematics that we are used to today. This helps us gain a wider view of the possibilities and contingencies of mathematical practice that contemporary imagination tends to suppress. In turn, this will provide us with a better, more complete understanding of the landscape that the title "mathematics" subsumes.

■ ■ ■ ■ ■ ■ ■ ■

CHAPTER 1

Histories of Philosophies of Mathematics

THIS CHAPTER SETS this book against the background of canonical philosophies of mathematics. It will present several historical narratives that follow some major trends in the philosophy of mathematics. All these narratives are going to be highly selective and superficial. I allow myself this bird's-eye view precisely because these histories are presented in parallel, emphasizing the partiality of each particular narrative. But this does not mean that together they presume to exhaust the historico-philosophical landscape. There's much more that can be added to these stories, and many ways in which they can be retold.

I should emphasize that these histories do not presume to explain or analyze the philosophical positions that they bring up. The point is rather to highlight some issues at stake in the debate among philosophers—issues that were not always explicitly highlighted by those philosophers. I will therefore not go into the details of each philosophical position or even provide a decent survey of their arguments. Doing that would force me to deviate too far from my main argument and reproduce some well-known discussions that are covered in many other introductory works (among the most recent are Bostock 2009; Celluci 2007; Friend 2007; George and Velleman 2001; Hacking 2014; Murawski 2010; Shapiro 2005; and the Stanford and Routledge online encyclopedias of philosophy). Since I only want to thread together the various debates to make salient some questions that underlie philosophical debates, many deep philosophical points that were very important to the various quoted philosophers will be glossed over very quickly and superficially, so as to pick up the problems that are important for the argument of this book.

More specifically, instead of philosophical questions such as "Are mathematical objects real?," "Is mathematics reducible to logic, or intuitions, or axioms?," or "Is mathematics a priori?," I will narrate the history of some philosophical discussions as revolving around metaphilosophical tensions: tensions between natural order and conceptual freedom; between mathematics as originally constitutive versus reducible to reason and/or nature; between reining in conceptual monsters and attempting to let them roam free; and between different ways to distribute authority over mathematical norms and standards.

History 1: On What There Is, Which Is a Tension between Natural Order and Conceptual Freedom

Our first narrative of philosophies of mathematics will start from W. V. Quine's (1948) "On What There Is." This paper raises the question of whether mathematical entities really exist or are figments of the imagination. This question relates to a common sentiment that says that mathematics is good because it provides good descriptions or predictions of empirical or idealized realities.

Quine's paper draws an analogy between, on the one hand, scholastic realism, conceptualism, and nominalism and, on the other hand, modern logicism, intuitionism, and formalism respectively (we're going to introduce the former and latter in a few paragraphs). In order to tease out the tension between natural order and conceptual freedom in the context of mathematics, we'll need to take a quick (and superficial) look at these schools of thought.

According to Quine's analogy, medieval realists and twentieth-century logicists were committed to the existence of all kinds of abstract objects; medieval conceptualists and modern intuitionists were committed only to those that could be admitted via a restricted mental construction; and nominalists and formalists were committed only to names and inscribed marks, without requiring objects to which these names and marks would refer beyond specific examples.

Quine's essay marks a pivotal moment in the history of twentieth-century philosophy of mathematics—so much so that Hilary Putnam stated that in the analytic tradition "[i]t was Quine who single-handedly made Ontology a respectable subject" (2004, 78–79). Quine's historical narrative came up at a time when logicism, intuitionism, and formalism have exhausted much of their drive as formative philosophical projects, and became absorbed into more technical work in logic. His pivotal statement set the scene for the contemporary realist-nominalist discussion that would become a leading concern in contemporary philosophy of mathematics.

To see what Quine's historiography entails, let's pause briefly to consider one dimension of the scholastic debate. But note that reducing a debate that spanned hundreds of years necessarily verges on a caricature and involves gross overgeneralizations. A properly detailed historical account of the fluctuating debate concerning universals in scholastic thought is provided by de Libera (1996).

Scholastic realism involved an elaborate scheme of entities that mediated human perception and divine knowledge, exceeding the bounds of time and space. It depended on a complex division of labor between the objects, the senses, the passive and active human intellect (which mediate between the senses and mental concepts), and divine illumination (that provides nonsensory input and direction). These elements combined to give rise to subjective and objective concepts (those constructed by a single individual and those necessarily shared by all) and to individualized and absolute essences (the essence of an object as understood by an individual and as it really is from a divine point of view). Universal concepts in general (for example, good, blue), and mathematical concepts in particular (numbers, geometrical forms), were to be abstracted from the sensed world by human intellects and divine illumination, and held together objectively across different minds and different individual instances by their common essences. This architecture forced the work of God and man into a rational Aristotelian hierarchical and purposeful frame integrated with a Platonic reification of abstract concepts as independent and eternal.

Nominalists reacted against this elaborate divinely rational world by doing away with many of the pillars that held the realist conceptual architecture together. Instead of abstractions of the essences from similar individuals, concepts or common names became just means of grouping individuals together. No universal concepts or essences held these different names or groups together—they were individual or collective engagements with the world. Divine will was no longer tied down to the heavy ontological burden of Aristotelian hierarchical and purposeful classifications and to Platonic eternal forms. The link between man and God depended less on reason and more on faith. Focusing on the cultural impact of this position (specifically, Ockham's fourteenth-century position), Sheila Delany suggests that

> Since God is bound by neither natural law nor his own promises, the universe becomes profoundly contingent and, in an absolute sense, unpredictable. Neither nature, society, nor the human mind are necessarily permanent or static in their structure; all are open to change and plurality, none can be fully understood by reference to an abstract *a priori* scheme. (Delany 1990, 48)

Divine and human wills became much more autonomous with respect to the nature of the created world. The cost, however, was that philosophical guarantees of communication, correct representation of the world, and a general sense of purpose could be lost.

Fitting this division with the early twentieth-century foundational trio of logicism, intuitionism, and formalism is tricky. First one has to bring in conceptualism to complete the scholastic duo into a triad (since the scholastic division was never as clearly defined as contemporary presentations pretend it to be, some of the more moderate forms of nominalism, which allow the mind to abstract some essence-ersatz from individuals, are termed "conceptualist"). Even then, if we consider the big picture, the early twentieth-century foundational positions don't have too much in common with the scholastic debate.

But Quine focused on quantification as a form of ontological commitment, not on the entire scholastic construction. Quantification is our use of the term "all ... " or "there are ... " in our scientific language. For Quine, using these terms meant that we acknowledge the existence of the entities that they refer to. So if we say "all numbers are ...," we commit to the existence of numbers. Now, if we restrict our attention to quantification alone, Quine's analogy works to an extent. Let's review this analogy.

Frege's and Russell's "logicism" advocated the reduction of mathematics to logic and pure reason. This seemed to be a tenable project, because by the end of the nineteenth century logic had become much more rich and expressive than it had ever been before. Their attempt was to present numbers and other mathematical entities as sets subject to the logical laws of some basic set theory. Logicists, as commonly interpreted, do quantify over abstract terms such as "sets," "numbers," and "geometrical shapes," as did scholastic realists (assuming we can retroject the term "quantify" a few hundred years into the past), and considered them to be real existing entities. As the early Bertrand Russell put it (and later retracted):

> All knowledge must be recognition, on pain of being mere delusion; Arithmetic must be discovered in just the same way in which Columbus discovered the West Indies, and we no more created numbers than he created the Indians. The number 2 is not purely mental, but is an entity which may be thought *of.* Whatever can be thought of has being, and its being is a precondition, not a result, of its being thought of, since it certainly does not exist in the thought which thinks of it. (Russell 1938, ch. 51, §427)

Brouwer's "intuitionism" (preceded in some ways by such figures as Poincaré and Kronecker) questioned any mathematics that could not be finitely constructed starting with counting a sequence of moments (in a Kant-like framework of temporality, to be discussed in the next section). In Brouwer's view, actual infinities could not be contained by such constructions and were therefore rejected.

Nonconstructive proofs were suspect as well. For instance, it's clear that *either* $\sqrt{2}^{\sqrt{2}}$ is an example of two irrational numbers yielding a rational power, *or*, if their power turns out to be irrational, then $(\sqrt{2}^{\sqrt{2}})^{\sqrt{2}} = (\sqrt{2})^{\sqrt{2} \times \sqrt{2}} = (\sqrt{2})^2 = 2$ is an example of two irrational numbers yielding a rational power. But as long as we cannot decide which alternative holds, we have failed to construct a rational power of two irrational numbers, and so radical intuitionists would say that we have not proved the existence of two irrationals yielding a rational power. This rejection of classical concepts and arguments required a thorough review of a lot of classical mathematics.

Intuitionists are harder to fit into Quine's analogy. They quantify over mental constructions based on the elementary intuition of temporal succession (counting the moments 1, 2, 3, ...), not over mere common (nominalist) names or over abstractions of fully fledged (realist) essences. One could therefore say that intuitionist and conceptualist objects exist in the mind. But what "mind" means, and how things come to exist in the mind works rather differently in the scholastic and intuitionist systems. The conceptualist scholastic mind collects individuals under concepts by analogies. The intuitionist mind constructs on its own by following the basic internal experience of the advance of time.

Hilbert's "formalism" tried to see mathematics as analyzing which combinations of signs can be obtained when following strict syntactic rules, without pretending to anchor these signs to any reference or meaning (think for example about algebraic equations with rules for simplifying them, but without deciding what kind of numbers the unknowns may stand for, or if they stand for numbers at all). But in Hilbert's system, such formal languages of signs and rules were to be analyzed and compared inside some system (a meta-language), and Hilbert determined this system to be the elementary core of finite arithmetic subject to constructive and logical restrictions that would be legitimate in the eyes of all philosophical parties involved.

Formalists make things more complicated for Quine's analogy, because of their double articulation of mathematical discourse into the language of formal proofs (meaningless signs following syntactic rules) and the meta-language that applies finite, constructive reasoning to the first-level languages.

Indeed, the formalist meta-language quantifies over numbers (it counts signs and makes claims that depend on the length of bits of text), but Poincaré and Hilbert were at odds as to whether the numbers of the meta-language were mere collective names of empirical marks on paper or full-blown mathematical mental constructions, especially where induction was used in the meta-language to make general statements about proofs (Ewald 1996, 1021–51).

As for the level of formal languages, in which mathematical proofs were written, according to the Hilbertian view these languages were collections of marks and rules that may discard any denotation. This is his famous statement that "point," "line," and "plane" in Euclidean geometry may as well be replaced by "tables," "chairs," and "beer mugs"; what matters is only the rules of the language (Euclidean axioms and postulates, as well as those that Euclid used implicitly), and we do not decide in advance what the Euclidean terms designate. (Hilbert actually constructed different kinds of geometry by offering different designations to the same Euclidean terms, so as to make them obey different sets of axioms.) The marks of formal language are therefore not names that group individuals together, as we would expect following Quine's analogy between formalists and nominalists.

So the analogy between the scholastic and early twentieth-century trichotomies is shaky at best. To find a more adequate modern version of scholastic nominalism, we should probably try John Stuart Mill, who stated that

> All numbers must be numbers of something: there are no such things as numbers in the abstract. *Ten* must mean ten bodies, or ten sounds, or ten beatings of the pulse. But though numbers must be numbers of something, they may be numbers of anything. (Mill 1843, book II, ch. 6, §2)

This means that when we say that "2 + 3 = 5," we make an empirical statement, but one that is collectively true for apples, chairs, and so on. It seems therefore that when Mill quantifies over numbers, he quantifies, like scholastic nominalists and like Hilbert in his meta-language treatment of numbers, over collections of individual instances.

If we proceed in time, and try to carry Quine's analogy over to contemporary realists and nominalists, we might find more substantial analogies. These analogies are not restricted to a concern with quantification. They even go beyond the borrowing of scholastic terms for realism and nominalism such as *in rebus* (existing in empirical things) and *ante rem* (existing a priori, independently of empirical things; see Reck and Price 2000 for details). I believe that the most substantial analogy between scholastics and modern philosophers of mathematics concerns the commitment to natural order versus conceptual freedom.

The following paragraph from Burgess and Rosen (1997, 241)—a critical study of contemporary nominalism in mathematics—is telling:

> the most fashionable figures in the history and sociology and anthropology of science deny not only that there is a ready-made theory of the world, but even there is any ready-made world. They maintain not just that theories about life and matter and number are constructs of human history and society and culture, but that number and matter and life themselves are such constructs [... and] that mathematical and physical and

biological facts, being created by us when we create mathematical and physical and biological theories, cannot impose any prior constraint on how we go about shaping those theories, leaving only constraints from our side—assumed to be social and political and economic—rather than the world's side.

The drums of the *Science Wars* can be clearly heard beating in the background, and contemporary forms of realism still seem concerned with the risk of relativism costing us our grip of reality. They sometimes even carry religious overtones. Mark Steiner's view (1998) is that some sort of anthropomorphic design is required to make the universe so highly accessible to human mathematics, and that if we disagree, then "[n]ominalism, like atheism, and for similar reasons, is a philosophical position that recommends itself to many modern philosophers" (2001, 73). Putnam's *Ethics without Ontology* (2004) is worried that losing grip of reality in the context of hard sciences is related to a dangerous descent into ethical relativity.

As in the scholastic realism debate, what is at stake is not only ontological commitments expressed by quantification but also the binding of human and divine reason to the reality of (Aristotelian or modern) science. But can we also fit into this analogy the aspect of medieval nominalism that set to liberate the human mind and will from some preordained conception of the created world?

Hellman (1998) jokingly reads Burgess and Rosen as accusing nominalists of Maoism—that is, of attempting to instigate a cultural revolution in mathematics. This revolution would seek to replace the mathematical language of science by other languages, which do not quantify over mathematical entities. Hartry Field (1980) is a paradigmatic example here. First, Field generates an alternative physical language that involves no mathematical entities (based on spatial locations and their observable qualitative relations). Then he shows that adding mathematics as a representational aid for this strictly physical language is "conservative"—that is, no conclusions drawn from the mathematical expansion of the language exceed those that can be drawn from the physical language itself. The advantage of the mathematical extensions is simply that it facilitates such derivations.

This so-called hard-road nominalism (I leave easy-road nominalism aside for now) may be read as a hygienic project: remove all mathematical language from physics, and show that letting it back in won't send us jumping to unwarranted conclusions. This project ensures our grip on reality by setting aside as contingent and harmless any mathematical shortcut involving entities that are not tangibly real. This kind of project is accepted by Burgess and Rosen, as well as Quine, as a means for figuring out what's necessary and what's contingent in the way we theorize the world. For Burgess and Rosen this is "a way of imaging what the science of alien intelligences might be like" (1997, 243), whereas Quine invited us to "see how, or to what degree, natural science may be rendered independent of platonistic [here, realist] mathematics; but let us also pursue mathematics and delve into its platonistic foundations" (1948, 38).

Given this reading of contemporary nominalism, we can go back to the first half of the twentieth century in order to bind the new nominalist logic of the fourteenth century and contemporary nominalist philosophy of mathematics. Our link is logical positivism, a project that sought to separate empirical claims from analytic claims (the latter being true by definitions and logical consequence). Mathematical statements were thus split into two kinds: to observe by counting that 3 apples and 2 pears make 5 fruit was an empirical observation; to claim that $3 + 2 = 5$ based on the definition of number and the associated rules was a purely analytic mathematical claim. Carnap's tolerance principle demanded that all analytic approaches be given equal opportunity to advance science. Those that work well with empirical science are to be retained, and those that do not can be discarded.

Hilbert's formalism can also be read in a similar vein. Finitary constructive reasoning (answering questions such as "Does this formula follow from that according to those given rules?"), which reigns over manipulations of marks on paper, is a physically embedded activity, and is therefore part of the real world (think also of computer-based verification—which discards the purported transcendence of human reasoning). Any mathematical language that expands this logic and is expressible in finite strings following finitely verifiable rules of derivation would be welcome into formalist mathematics regardless of what (and if) it purported to represent. Hilbert's only qualification was that the new addition wouldn't lead us to false conclusions about the finite constructive arithmetic we had started with. Hilbert hoped that this form of conservatism would be provable by means of the meta-mathematical finitary constructive logic.

Like medieval nominalism, these forms of modern mathematical nominalism (hard road, logical positivism, formalism) promote conceptual and linguistic freedom while drawing a clear line dividing this freedom from hard physical fact. Contemporary nominalism, however, cannot rely on medieval faith, and tries to prove that this freedom to reason does not generate false claims concerning the actual world. But none of these contemporary forms of nominalism is generally accepted as a successful endeavor. And so, many philosophers still believe that only a holist realism where all terms refer to existing (physical or ideal) entities can protect us from the vagaries of errant schoolmen and fashionable sociologists.

But recall here the mathematical vignette from the introduction. We saw that the Black-Scholes formula is of value even where it strays from the empirical path, even when it allows us to draw empirically false conclusions or has no clear empirical counterpart. This does not, however, make the formula arbitrary or meaningless—it is useful precisely because practitioners successfully endow it with meaning and make it relevant for the actual world of trading.

History 2: The Kantian Matrix, Which Grants Mathematics a Constitutive Intermediary Epistemological Position

Following the echoes of nominalist scholastic thought, French luminaries pointed to a gap between real mathematics and its abstract counterpart. Condillac, for example, found that numbers began from concrete representations of objects by fingers and so on, but then, as numbers were abstracted, they lost their footing in objects and were perceived "in the names that have become the signs of the numbers" (Condillac, *Langue des Calculs*, quoted in Schubring 2005, 260). Others, like d'Alembert, insisted that the gap between real and nominal mathematics had to be bridgeable:

[W]e arrive through the continual generalization of our ideas at the principal part of mathematics.... [But then mathematics] retraces its steps, reconstitutes anew its perceptions themselves, and, little by little and by degrees, produces from them the concrete beings that are the immediate and direct objects of our sensations. (d'Alembert 1995, 20–21)

Kant's crucial contribution to the philosophy of mathematics (arguing not directly with French luminaries, but with a tradition going back to the rationalist Descartes and empiricist Hume) was to revise his contemporaries' conception of the gap between empiricist and rationalist views by placing mathematics in an intermediary position. This maneuver was organized by means of the Kantian matrix.

This matrix classifies judgments or assertions (there's a distinction here, but I won't pick on this nit) according to two axes: analytic/synthetic and a priori/a posteriori.

- An assertion is "analytic" if the predicate (or what is claimed) is part of the concept-subject (or that of which it is claimed). That a triangle has three sides is analytic, because having three sides is contained in the definition of a triangle.
- An assertion is "synthetic" if the predicate adds something to what the concept of the subject already contains. To claim that some triangle has a side whose length is 10 inches is synthetic, because it does not follow from the concept of triangle.

  (There's a problem with what it means for the predicate to be properly included in a concept. Does the predicate have to be literally included, or can it be derivable from the concept by logical reasoning? For Kant, this wasn't much of an issue, because what was considered as logical derivation at the time, based on Aristotelian syllogisms, had a very limited reach. But as the logic of functions, relations, and classes evolved, the issue became more pertinent.)
- An assertion is "a priori" if it requires no empirical experience to be made. For example, to know that anything is identical with itself, at least according to Kant, no experience is required.
- An assertion is "a posteriori" if it is based on empirical experience. To assert that the sun is shining right now depends on my observation of the sky, or some other, less direct observation.

One might expect the axes of a priori/a posteriori and analytic/synthetic to overlap. Indeed, if an assertion does not depend on experience, then we'd expect its truth to follow from definitions and logic, and vice versa. But Kant finds exceptions. While he thinks that analytic a posteriori assertions are impossible (if the predicate is included in the concept, there's no need to resort to experience), he finds that synthetic a priori assertions are typical of mathematics.

For example, consider the assertion that the sum of angles in a triangle is two right angles. The Euclidean proof requires a construction: draw a parallel to one of the sides through the opposite vertex. This construction is not part of the concept of a triangle, but is necessary to establish the assertion. The assertion that the sum of angles is two right angles therefore exceeds the concept of a triangle. It is synthetic.

On the other hand, no external empirical experience is required for the Euclidean proof. One can run through the proof with one's eyes closed in a sensory deprivation tank. This synthetic assertion is therefore a priori. The synthetic a priori is thus a realm of knowledge that does not depend on external observation, but is still richer than pure logic. It includes the pure forms of time and space on the surface of which mathematical reasoning takes place. The a priori experience of this realm is termed by Kant "pure intuition" (and has nothing to do with the use of the term "intuition" as some sort of vague premonition—this is rather the nonsensory intuition in which intuitionists carry out their constructions).

Much of the subsequent philosophy of mathematics can be read as a reaction to this division of labor between reason's a priori analytic logic, the senses' empirical synthetic a posteriori observations, and Kant's synthetic a priori middle ground. The juncture here concerns the place of mathematics in the greater order of knowledge: is it a distinguished, necessary, and formative middle ground, or is it reducible to reason or empirical observation?

A first step in the evolution of post-Kantian ideas was to ask whether, even if we accept Kant's architecture, all mathematics is synthetic a priori. Even Kant's most faithful immediate successors pointed out that some mathematics was nevertheless analytic. I'm not referring here to such obviously analytic assertions as "a triangle is a triangle" or "a triangle has three sides," but to claims that involve infinities and infinitesimals. Our source here is the Kantian philosopher Jakob Friedrich Fries (1773–1843).

According to Fries, there are two kinds of infinities. The first fits into the Kantian synthetic a priori. It is the infinite as "indefinite." It follows Kant's solution to the antinomies of pure reason: our synthetic a priori experience of space shows us that we can always go farther in any direction, but we cannot thereby experience or deduce any complete infinity of the world. The process of indefinite increase or diminution is therefore the synthetic a priori version of infinity, and is admissible in a Kantian mathematical worldview. It is expressed, for example, in convergent infinite series and notions of limit.

But reason itself can come up with a concept of complete infinities or infinitesimals, as witnessed by the many mathematicians who appealed to such infinities before Kant. Now Kant argued that thoughts without content (that is, devoid of any pure or empirical experience) are empty. The concept may be had, but it will not have any real counterpart and will have no scientific purport.

Fries, influenced by the mathematician Carl Friedrich Hindenburg, sought to find a place for reason's complete infinities. So he distinguished between syntactic operations (combinations of signs that follow arbitrary rules) and arithmetic operations (operations that deal with magnitudes in the framework of Kant's synthetic a priori). The former depend on combinatorial order, and the latter on the classical axioms of magnitudes and ratios. The syntactic approach is therefore independent of the arithmetic one (Fries 1822, 68).

Due to the Kantian exclusion of complete infinity from intuition, operations with complete infinities and infinitesimals were to be considered as strictly syntactic (Fries 1822, 280, 294). Mathematical operations with infinities, like those resorted to by such mathematicians as Wallis, Euler, and Fontenelle (Boyer 1959), had nothing to do with empirical experience, and so came down to games of reason. Any empirically viable claim involving complete infinities in the spatio-temporal world of measurable magnitudes would have to be reduced to expressions of indefinite increase or decrease.

The important point here is that one admits a form of mathematics that is analytic and a priori, depending on empty reason—concepts that lack a corresponding pure or empirical intuition beyond their articulations in signs. According to Fries, these are figments of reason's creative imagination. Just because they're analytic and a priori, it doesn't follow that they are real.

Now this approach was not satisfactory for leading nineteenth-century philosophers. Leaving reason to hover irrelevantly in the background did not seem appropriate. Much of German idealism was about identifying reason with being (Schelling; see chapter 7) or even claiming that it forms being according to a teleological plan (Hegel).

This trend did not die away with German idealism, but did assume less extravagant forms. By the time of Gottlob Frege (1848–1925), and partly due to his own efforts, logic could be made rich enough to describe a wide variety of arithmetic structures and derivations, and so Frege attempted to draw arithmetic (if not all of mathematics) back into the analytic a priori. But this was not to suggest that mathematics was a figment of imagination or a game of reason. For Frege, mathematics was real:

> The thought we have expressed in the Pythagorean theorem is timelessly true, true independently of whether anyone takes it to be true. It needs no owner. It is not true only from the time when it is discovered; just as a planet, even before anyone saw it, was in interaction with other planets. (Frege 1984, 363)

I won't get into the intricate discussion over the precise nature of Frege's supposed realism. The point is that Frege did not allow mathematics to be tied to any form of experience, not even a necessary foundation for all human experience such as Kant's synthetic a priori. Logicism wanted to push mathematics back into the realm of the analytic a priori, without thereby giving up its status as objective truth about the world.

Another branch in the evolution of Kantian ideas is intuitionism, referring to Kant's pure intuition, the experience of space and time that precedes empirical observation. Mathematicians such as Poincaré and Brouwer sought to retain mathematics' status as a distinguished "in between." For Brouwer this meant that reason's complete infinities had to be given up. The indefinite sequence of integers is to be kept, but higher infinities were to be excluded, even at the cost of a wholesale revision of the mathematics of the continuum of real numbers.

Around the same time, Hermann Cohen and Ernst Cassirer preferred to follow Kant by adhering to something like the synthetic a priori, but chose to enrich it, rather than give up higher infinitary analysis. To maintain mathematics' distinguished intermediary position as well as its richness, Cohen (1883; see chapter 7) considered the infinitesimal as the distinguished synthesizing foundation of mathematics; Cassirer (1910) opted for the modern notions of function and relation, rendering the focus on mathematical "objects" obsolete. But the most interesting innovation of Cohen and Cassirer is the historicity and contingency they introduced into the synthesizing middle ground. This middle ground is no longer a given foundation of human knowledge to be deduced by solving apparent paradoxes. What serves as a synthetic a priori background to our reasoning is a historically and culturally formed intermediary between reason and reality—one possible manner to found human worldviews.

Hilbert's formalism, on the other hand, seems most faithful to Fries's articulation of arithmetic versus syntax. Hilbert's meta-language is a minimal mathematics that purports to respect the demands of intuitionists, logicists, and those who want to reduce mathematics to a nominalist abstraction from empirical finite counting. (I don't think he was too interested in respecting the Kantian synthetic a priori, as he was keen on starting from a foundation that no one in his philosophical environment would question.) The rest of mathematics was not to be rejected; it was to be upheld as purely syntactic. However, to make sure that the purely syntactic mathematical elaborations that we introduce into our reasoning do not shake the consensual foundation, consistency must be guaranteed, and it must be proven by the means available within this consensual core. Gödel's second incompleteness theorem shattered this hope.

To tie things to the previous section, note that logical positivism sought to do away with the distinguished in-between position of mathematics. It allowed only empirical synthetic a posteriori assertions, and logical analytic a priori ones. The cost was, as explained earlier, to split mathematical claims into two aspects: an empirical descriptive aspect and a formal syntactic one. Following on this tradition, contemporary realism and nominalism don't settle for this dualism; they demand that we choose sides.

Our first history brought up the tension between an understanding of mathematics as committed to natural order and its view as conceptually free. The current history shows a tension between placing mathematics in a distinguished intermediary foundational position versus its absorption into the naturally ordered world and/or into free thought. The underlying question is whether mathematics is so special and foundational that it deserves its own ontologico-epistemological domain.

Going back to our vignette, we saw that the Black-Scholes formula is not strictly descriptive, nor is it an arbitrary, free construct. However, it's hardly describable as belonging to a distinguished a priori foundation or necessary truth. An account of mathematical practice must find a different articulation in order to explain its force.

History 3: Monster Barring, Monster Taming, and Living with Mathematical Monsters

Our starting point here is Aristotle's prohibition against *metabasis eis allo genos*—the transfer of reasoning between different kinds of entities. "[W]e cannot in demonstrating pass from one genus to another. We cannot, for instance, prove geometrical truths by arithmetic" (Posterior Analytics I.7 75a38–39, G.R.G. Mure's translation). According to Aristotle, there's geometry dealing with continuous lines, and there's arithmetic dealing with discrete numbers, and we should not mix the two together or bad things might happen. Much of the subsequent history of mathematics and physics is often read as the story of gradually bridging the gap toward a unified mathematical science.

But this is a problematic narrative. It is obvious that spatial measurement and enumeration have developed together. When one measures land or cloth for taxation or trade, one uses numbers. When one reasons about numbers, spatial organization (triangular numbers, square numbers, rectangular numbers) plays a significant role. Geometry and arithmetic grew together, and were then forced apart by ancient highbrow scholars.

The received narrative says that the Pythagoreans discovered that some ratios between lines (such as the side and diagonal of a square) were "incommensurable"—that is, their ratio could not be expressed as a ratio between whole numbers. The reaction to this crisis was the extravagant overkill captured in Aristotle's prohibition: remove all arithmetic from geometric discourse. Instead of stating that when some lines are given numerical values, others might be expressible numerically only by approximation (which is the practical thing to do), classical Greek geometry was exhaustively recast with no reference to numbers. The monster of incommensurability or irrationality was barred from arithmetic by caging it in the realm of geometry.

But even in classical Greece, practical people went on with the daily business of mixing geometry and arithmetic (Asper 2008). It was only a small elite that practiced pure geometry. This elite constituted a distributed network of players writing in a practically secret language—not in the sense of a cipher, but in the sense of a highly codified lexicon, syntax, and logic that could not be imitated without proper initiation (Netz 1999; Latour 2008). Being an esoteric language mastered by an elite minority surely helped geometry become venerated in Plato's academy as a gateway to some higher truth beyond the grasp of land surveyors and accountants. This scholarly approach barred not only the monster of irrational numbers but also the unjustified, inconsistent, and imprecise practices of common folk.

Soon enough, however, monsters crept up again (from the East!). The mathematics of South and West Asia merged and brought irrational numbers—square roots treated as numbers rather than as geometric magnitudes—to the Arabic heirs of classical Greek science. As those monstrous algebraic entities grew more common and useful, they had to be integrated, rather than barred.

This was a project that started with some of the earliest surviving Arabic algebras (al-Khwarizmi and abu Kamil), which provided Euclidean representations and proofs for the algorithms solving quadratic equations (Rashed 1994). In Renaissance Italy, algebra grew wilder and allowed even more suspect entities, such as subtractions of larger numbers from smaller ones and the roots of their results. The epitome of the efforts to geometrize Italian algebra, including such monsters as roots of negative numbers, belongs to Bombelli's *L'Algebra* (see chapter 6, figure 6.4 and discussion). It was soon superseded by the Cartesian analytic method at the brink of early modernity.

We see here that plural practical mathematics, which lives with monsters, and codified scholarly mathematics, which seeks to bar monsters, have a third alternative. We'll call it monster taming: the translation of new and suspect entities into the language of an established foundation. The tension between the three strategies is one that mathematics continued to engage with throughout its future.

With the onset of early modern science, a new monster reared its head: the infinitesimal. Infinities and infinitesimals became constitutive elements of proto-calculus and calculus. The Jesuits, Thomas Hobbes, and George Berkeley all fought to banish them from mathematics due to their dangerous hybrid properties and inconsistencies (Alexander 2014; Berkeley 1734). Observables and the dicta of authorities were to be the sole basis for mathematics. Infinitesimals were to be barred.

Other mathematicians, like Euler, were happy playing with the monster. They knew it involved contradictions, but they engaged with it nonetheless (Jahnke 2003). This does not mean that they were happy with inconsistencies. In fact, when Euler calculated the sum of the divergent series $\sum_{n=0}^{\infty}(-1)^n n!$, he was comfortable with his result precisely because it was arrived at consistently in several different ways (Sandifer 2007, ch. 31). Contradictions were not welcome, but they were avoided by picking methods in context, rather than by postulating a priori rules.

The third approach, that of monster taming, was explored at the time as well. Some of the earliest to use infinitarian arguments (for example, Fermat, Newton, MacLaurin) either argued for the possibility of, or actually practiced, a reduction of infinitesimal arguments to proofs by contradiction following the Archimedean method of exhaustion. Instead of using infinitesimals to calculate the area of curvilinear shapes, they used improving rectilinear approximations to show that any value lower (respectively, higher) than the calculated area is exceeded by the area of a rectilinear shape circumscribed in the original shape (respectively, exceeds the area of a rectilinear shape circumscribing the original shape).

Later, Carnot (1813; see also Schubring 2005) offered a whole catalogue of attempts to reinterpret infinitesimals by means of established mathematical entities. These attempts boil down to the method of exhaustion from the previous paragraph, limits of ratios of evanescent quantities (based on Newton's first and last ratios), and two algebraic methods: Lagrange's identification of derivatives with coefficients in power series instead of ratios of infinitesimal differentials, and an original attempt to view infinitesimals as independent corrective variables (more detail is available in the second section in chapter 6). Rather than choose, Carnot advocated a pluralistic approach. Monster taming by translation does not necessarily mean a unique translation. Even Cauchy, who is usually understood as having reduced infinitesimals to limits, is most convincingly read as striking a compromise between limits and infinitesimals (Schubring 2005).

But these plural compromises didn't last, and by the end of the nineteenth century infinitesimals were reduced to sequences of vanishing numbers. Note the foundational twist here: as geometry, with its new non-Euclidean and projective varieties, lost its age-old stability, arithmetic became the leading candidate for monster taming. The project of arithmetization spanning from Bolzano to Weierstrass and Dedekind culminated in the attempt to reduce all mathematics to the sequence of natural numbers and its subsets.

But the tamed infinitesimals and irrationals of arithmetic gave rise to new monsters, which in turn invoked a counterinsurgency of monster barring. The monsters were Cantor's infinite sets and their paradoxes, bound together with highly nonanalytic functions (continuous functions without derivatives, space filling curves, and so on). Poincaré protested: "Formerly when a new function was invented, it was in view of some practical end. To-day they are invented on purpose to show our ancestors' reasonings at fault, and we shall never get anything more out of them" (1914, 125).

The most famous attempt at barring these monsters is Brouwer's intuitionism. It reduced mathematics to a constructive core, excluding the law of excluded middle and complete infinities. The work of Lebesgue, Borel, and Baire on descriptive set theory had followed a related path. According to this approach, sets of numbers could not be trusted unless they belonged to a controllable hierarchy of increasingly complex constructions (Cavaillès 1994). Discussions of constructability turned the axiom of choice (saying, roughly, that I can simultaneously choose one element from each set in an infinite collection of sets) from something too obvious to even notice into an independent, questionable axiom. But these attempts at monster barring proved either too restrictive or monstrous in their own ways.

Hilbert's formalism offered a new form of monster taming. In his architecture, monsters could roam free, as long as their mathematical habitats could be axiomatized in a contradiction-free manner. We don't need to choose between mathematics with or without infinity, with or without the axiom of choice, with or without nonconstructible sets. Each kind of theory can be developed independently based on its own consistent axioms and without conflict with alternative theories. But we shouldn't jump to the conclusion that Hilbert's monster taming has won over the constructivists' monster barring. More and more mathematicians are interested in constructive versions of existing theories—not due to ontological concerns, but because of algorithmic applications.

It seems that modern mathematics won't stand for living with untamed monsters. But this is not precisely the case. Some mathematicians interested in the foundations of category theory and set theory are concerned that their conjunction does not have a well-articulated, consistent axiomatic foundation (though this is not a universally endorsed position). As Pierre Cartier stated in an interview:

> Nowadays, one of the most interesting points in mathematics is that, although all categorical reasonings are formally contradictory, we use them and we never make a mistake.... [A] revolution of the foundations similar to what Cauchy and Weierstrass did for analysis is still to arrive (Fresán 2009, 33; see also the abstract in Krömer et al. 2009, 472–74).

But Cartier still claims here that contradictions must be resolved. This is not as obvious as it sounds. Concerning the contradiction that Russell discovered in Frege's set theory, Wittgenstein stated:

> If you based something on this system, I don't see that it would necessarily be detrimental if there were a contradiction in it, as long as this contradiction is just not used as a thoroughfare or circus [thus allowing to derive any statement whatsoever].... The only point would be: how to avoid *going through* the contradiction unawares (1976, 227).

While Turing reacted negatively to this assertion, contemporary paraconsistent logic (which allows contradictions without collapsing into triviality by rejecting the truth of some implications with a false premise) is precisely in line with Wittgenstein's advice on taming the contradiction.

It is important to mention here Imre Lakatos, who developed a beautiful dialectical conception of mathematical practice based on managing the encounter with monsters (his own term) in *Proofs and Refutations* (1976). According to this dialectic, a mathematical claim is made and justified. Then a "monster"—a so far unthinkable example— shakes the generality of the claim. The claim is then reformulated to fit the conceptual impact of the new example by various strategies.

This third narrative discussed a historic tension between three meta-mathematical approaches: monster barring (excluding problematic or contradictory entities from mathematics); monster taming (translating these entities into more acceptable mathematics); and living with monsters (acknowledging contradictions and avoiding them by pragmatic experience). Dominant philosophies of mathematics usually do not engage this problematic. They tend to assume that monsters have already been barred or tamed (a particularly challenging exception is Colyvan 2010).

But monsters are dangerous. The monstrous probabilistic models ignored by the Black-Scholes formula (noncontinuous prices, distributions with infinite moments) came back with a vengeance. According to some analysts, it was due to their very presence that option pricing formulas failed and led to markets crashes (Nassim Taleb's so-called black swans). Now this interpretation must be taken with a grain (or rather a handful) of salt. Indeed, markets have been crashing for centuries, long before Black-Scholes came to the fore. But trying to pretend pricing model monsters away may have contributed to the specific structural circumstances of the most recent crashes. Closing the door on monsters left a wide-open window for them to crawl back in …

History 4: Authority, or Who Gets to Decide What Mathematics Is About

Let's track back to the beginning of the last historical narrative. I followed Netz's claim that the creation of a codified elitist language helped bestow on elitist Greek mathematics the aura of access to a higher truth. Now this aura is not something *sui generis*. This impression had to be encouraged.

One of those who encouraged it was Plato. In the 7th book of the *Republic*, he wrote that "the knowledge at which geometry aims is knowledge of the eternal, and not of aught perishing and transient.... [G]eometry will draw the soul towards truth, and create the spirit of philosophy." While Plato found his contemporary standard of mathematical teaching unsatisfactory, he believed it "would be otherwise if the whole State became the director of these studies and gave honor to them; then disciples would want to come, and there would be continuous and earnest search, and discoveries would be made" (Plato 1973, 219–20). We see that for Plato mathematics is authorized by and for philosophy, and that the state is called upon to enforce this authority. But this does not mean that authority, philosophy, and mathematics necessarily go hand in hand.

The history in the previous section proceeded to the early modern concern with barring or living with infinitesimals. As Amir Alexander makes clear, this concern was highly overdetermined by concerns over authority. "In Italy it was the Jesuits who had led the charge against infinitesimals, as part of their efforts to reassert the authority of the catholic church" (Alexander 2014, 13). And later in England, when "the low mixed with the high, and boors such as Wallis were allowed in high society, what hope was there for court and king to establish their [Hobbesian absolute] authority?" (Alexander 2014, 7).

As for Berkeley versus Newton, things are even more explicit. The critique of Newton's method in *The Analyst* (1734) revolves around the supposed supremacy of mathematical reasoning over religious reasoning. Question 49 reads:

> Whether there be not really a *Philosophia prima*, a certain transcendental Science superior to and more extensive than Mathematics, which it might behove our modern Analysts rather to learn than despise?

And question 63 follows with:

> Whether Mathematicians, who are so delicate in religious Points, are strictly scrupulous in their own Science? Whether they do not submit to Authority, take things upon Trust, and believe Points inconceivable?

Whether they have not their Mysteries, and what is more, their Repugnancies and Contradictions?

Mathematical, political, and religious authorities were engaged in a conflict that surfaced around the question of infinitesimals. Philosophical critiques or praises of mathematical reasoning were thoroughly entangled with social alliances around this division of authority.

The concern with authority gains an interesting twist at the beginning of the nineteenth century. It's no longer about constitutive authority—who has the authority to dictate logico-mathematical terms—but about creative authority, namely, who creates mathematical knowledge.

German idealism, starting with Fichte, had a hard time accepting Kant's dichotomies (see chapter 7). They sought to unify a priori reason and a posteriori being. One way of doing that was to let reason and creative imagination generate knowledge not only when applied to given intuitions, but also by constituting it directly. In this version, geometry is not given by a transcendentally deduced synthetic a priori, but generated by the subject as free to form and construct its own world.

Fichte asserts that space and points are not sufficient for geometry to arise in human culture. We require one further ingredient that for him is tacitly assumed in the Euclidean system—*freedom*. We have to presuppose the free acting of the human being in order to move points around in space into a line: "Geometry arises through the free acting of my I, by moving the point into a line in space" (Wood 2012, 89, quoting the Zürich Wissenschaftslehre lecture).

This way of thinking finds its echoes in the discourse of mathematicians. Dedekind, for example, stated that turning a line of rational numbers into a continuous line of real numbers may depend on "a creation of new point individuals" (Ewald 1996, 772), and that the existence of an infinite sequence is proven by observing the objects in "my own realm of thought" (Ewald 1996, 806). Kronecker's famous comment about God creating only the natural numbers, leaving the rest to humans, is compatible with this sentiment, except that Kronecker uses it to disparage what he considered dubious human creations—the very creations that Dedeking exalts.

When we reach Brouwer, we find that "*mathematical contemplation arises in two phases as an act of the will in the service of the instinct for self preservation of the individual man*" (Ewald 1996, 1175). This refers to the abstraction of the sequence of natural numbers from temporal and causal attitudes. The language of Brouwer's passages is filled with the style and jargon of German idealism and its aftermath, projecting all the way into Husserl's phenomenology.

My point here is not to reiterate the previous history's concern with constructivism for the purpose of monster barring. Here we are dealing with the creative authority of man with respect to mathematics, and with a coalition of mathematicians and philosophers formed in order to assert the authority of man as creator. But we should note, at least in passing, another trend that surfaced at the same time—namely, the rise of statistics. Statistics is, by definition, the mathematics of the state, designed to allow governments to rule, forming a strong coalition between mathematics and political authorities.

While mainstream continental philosophy up until the mid-twentieth century tended to view man (and in the twentieth century, increasingly, women too) as self-creator, it was the analytic tradition that posited mathematics as a central concern. In this tradition, scientific realism has been gaining ground, partly relying on the "indispensability argument" attributed to Quine and Putnam: the claim that we should believe in the reality of mathematical entities because they play an indispensable part of contemporary science, which, for analytic philosophers, is our most successful attempt to understand and define reality.

This argument means that authority has been shifted from self-creating man and the investigations of philosophers to modern science. Mathematics gains its status not because of its introspection or foundations, but because of its indispensible services rendered to science. As Reichenbach put it, adherents of his philosophy

> refuse to recognize the authority of the philosopher who claims to know the truth from intuition, from insight into a world of ideas or into the nature of reason or the principles of being, or from whatever super-empirical source. There is no separate entrance to truth for philosophers. (Schilpp 1949, 310)

But there's a risk to mathematical autonomy in this approach. What happens if new science turns out to make less essential uses of mathematics? And what about the many branches of mathematics that are useless for empirical science? Do they become less objective or less true?

Naturalist philosophy of mathematics tries to counter this stance: "sets are just the sort of thing set theory describes; this is all there is to them; for questions about sets, set theory is the only relevant authority" (Maddy 2013, 61). The authority over mathematics belongs to mathematicians. No one else can do a better job than mathematicians in scrutinizing their work. But even here, Maddy's choice to respect the standards of mathematicians (rather than, say, those of astrologers) has to do with mathematical conformity with the overall modern scientific project (Maddy 2013, 350–51).

The issue at hand is therefore who has authority over mathematical standards: mathematicians, philosophers, scientists, statesmen? We can, of course, acknowledge that the answer is "all of the above." Only a sustainable coalition of players in the preceding fields can give rise to strong and viable mathematical institutions. But the philosophical debate seeks to strengthen the position of some players at the expense of others in order to facilitate certain coalitions and standards.

Note one more dimension of this debate. Haug and Taleb (2011), quoted in the introductory vignette, severely attack mathematical economists who pretend to assert authority over the economic practice of trading ("scientists lecturing birds on how to fly"). Here the authority dispute is not about mathematics, but about mathematics overstepping its boundaries, and interfering with economic practice.

The "Yes, Please!" Philosophy of Mathematics

If there's one thing that's common to all the philosophical currents surveyed in the preceding histories (but by no means to all philosophies of mathematics), it is that they make us choose. We're expected to decide that mathematical objects are either in nature or they're not; that mathematical statements are either synthetic or analytic; that monsters should be allowed into mathematics or that they be barred; that mathematicians should set their own standards, or that they should let philosophers, scientists, and agents of power trace their path.

But making us decide assumes that mathematics is monolithic enough to warrant a decision. And it's not. Mathematics is many things under a common name, which obey many conflicting norms and standards. We don't have to choose. The correct answer to the philosophical questions articulated in this chapter is an emphatic "all of the above!" In some specific contexts, one may have good reasons to persuade mathematicians or philosophers to prefer one choice over another, but if we want to think of mathematics as a whole, we need to acknowledge that the different philosophies of mathematics capture different aspects of mathematical practice (for a different, more analytic pluralist approach to mathematics see Friend 2014).

This kind of answer seems to turn philosophy of mathematics into a pointless effort; but this is also wrong. Given the plurality of mathematics, we still have to explain what mathematics does, how it works, and what makes different expressions of mathematics stand out in our scientific culture. We need to come up with an account that helps us figure out how we would like to relate to mathematics in our various schemes of things—but all that will be deferred to chapter 3.

Before we go there, the next chapter will demonstrate that mathematics can indeed be so many different things, even if we look at a particular branch of mathematics in a particular time and place. We will show what kind of plurality the philosophy of mathematics must embrace, if it is to be faithful to the phenomenon that it seeks to explicate.

CHAPTER 2

The New Entities of Abbacus and Renaissance Algebra

Abbacus and Renaissance Algebraists

This chapter will offer a historical narrative of some elements of the new algebra that was developed in the fourteenth to sixteenth centuries in northern Italy. Before we begin the story of mathematics, though, I'd like to say something about the mathematicians.

The mathematics we're considering emerged from a practical context. It was written by abbacus masters—teachers who ran arithmetic schools for merchant children. Despite their title, they had little to do with the abacus as instrument of calculation (this is why I follow Jens Høyrup's spelling here, based on the vernacular variant *abbaco*). Their teaching focused on decimal representation with its calculation algorithms (*algorism*) and practical problems of exchange, partnership, interest, measurement, and so on.

The teachers' status was parallel to other lower and mid-level liberal professionals, below that of jurists and medical doctors. Aside from teaching mathematics, these masters worked as engineers, surveyors, and accountants. They were paid either by local authorities or directly by the parents of their students (Ulivi 2002, 2008).

While algebra was not taught in their schools, some masters pursued algebra as a leisure activity. They also used algebra to gain renown and impress prospective clients (Høyrup 2008). Not all their textbooks include algebra (for example, Swetz 1987), but among the hundreds of surviving manuscripts listed in Van Egmond's catalogue (1980), there are quite a few dealing with algebra, including some that develop it creatively in a period that is usually supposed to be one of mathematical stagnation. Reviews of abbacist algebra and its sources are available in Franci and Rigatelli (1985) and Høyrup (2007).

In the sixteenth century, the vernacular culture of abbacus masters and the Latin humanist and university cultures began to merge. This is the time of the famous solution of cubic and quartic equations by Dal Ferro, Tartaglia, Cardano, Ferrari, and Bombelli. This is also the period of decline of abbacus schools and rise of the new Italian science. This chapter will try to follow some of the important mathematical developments of this period, and analyze them in terms of a "Yes, please!" philosophy of mathematics.

The Emergence of the Sign of the Unknown

Our starting point is the manuscript page in figure 2.1. This page recounts a problem about five men finding a purse. The first person says that if he were given the purse, then together with his own money he would have $2\frac{1}{2}$ times as much as all the others combined. The second says that if he were given the purse, he would have $3\frac{1}{3}$ times as much as the others. The others follow suit, ending with the fifth person stating that if he were given the purse, he would have $6\frac{1}{6}$ times as much as the others.

This problem has a long history. Problems with the same structure occur already in Greek sources. A version with an actual purse occurs in Mahavira's seventh-century Sanskrit arithmetic. The specific problem recounted earlier was stated by Leonardo Pisano, better known as Fibonacci, at the beginning of the thirteenth century, 250 years before Benedetto's rendition (Singmaster 2004, §7.R).

One striking feature of Benedetto's manuscript is the division of the page into calculations and running text. As Jens Høyrup noted (2010), the calculations occur on the left-hand side; only after they were done was the running text appended to the right. The calculations include elements that look very much like modern linear combinations of variables (figure 2.2). Comparing the calculations to the running text (figure 2.3), we see that the letter *b* in the calculation diagrams stands for *borsa* (purse), and the crossed *q*, a ligature for *qua*, stands for *quantità*

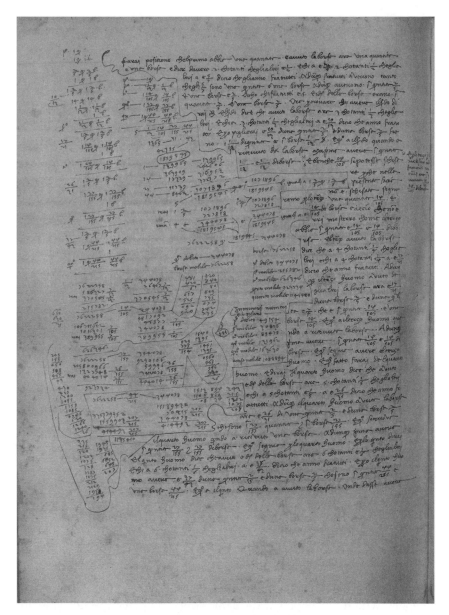

Figure 2.1: Manuscript page from Maestro Benedetto's fifteenth-century Florentine *Trattato de Praticha d'Arismetrica*. Copyright Biblioteca Comunale degli Intronati, Siena, L.IV.21, fol. 263v. 2015. 11, 24.

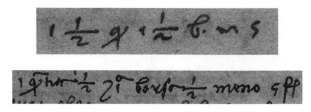

Figures 2.2 and 2.3: Details from Benedetto's *Trattato de Praticha d'Arismetrica*. Copyright Biblioteca Comunale degli Intronati, Siena, L.IV.21, fol. 264r. 2015. 11, 24. The shorthand (top) reads: "1½ q 1½ b. m 5"; the running text (bottom) reads: "1 *quantità* [and a] ½ and 1 *borsa* [and a] ½ less 5 fl."

(quantity). The former represents the amount of money in the purse, and the latter represents the quantity of money of the first person in the story.

The obvious question is: where does the practice of using letters and ligatures to represent unknown quantities come from? Historically, this is hard to track down. Since calculations were done either on dustboards or scrap paper, we have few surviving calculations schemes such as the one in Benedetto's manuscript (Høyrup 2010). But here we have an easy contextual answer. Indeed, the text in figure 2.3 ends with a ligature representing the monetary unit *Fiorino*.

This usage indicates that the sign of the unknown quantity did not have to be "invented." It was appropriated from economical practice. The *b* is not just an abbreviation of *borsa*; it serves as a sort of monetary denomination—the very amount of money that one finds in the purse. The same goes for the crossed *q*. Adding the two together is no different from adding different kinds of money in one line. Calculating with these signs does not require any new ideas.

The economic context may also explain the way that the calculation in figure 2.1 is set: a column of consecutive transactions applied to a given amount of money. This echoes the system of double-entry bookkeeping, which developed at the same time and place by the same abbacists (a strict relation between the two ways of writing is not proven conclusively, but the analogies are substantial; see Heeffer 2011).

So far, however, the discussion has only concerned the *form* of representation: the symbols used and the ways they are set on paper. What about the concept of an unknown or variable magnitude? Can we account for it from within the economic context of the Italian abbacus tradition?

In order to give an account of the emergence of a concept of an unknown magnitude, I picked out three quotations from two abbacus treatises. Each illuminates a slightly different aspect of how mathematical variability emerges from economic practice. Each quotation is rather trivial and unremarkable. But I think that together they point to where algebraic unknowns and variables come from.

The first is a "rule-of-three" exercise. Suppose you know that the price of A units of a product is C, and want to know the price of B units of the same product. The rule of three is the standard procedure that answers this question by taking the known price C, multiplying it by B, and dividing by A. Here's one variant from Piero della Francesca's *Trattato d'Abaco* (this is indeed the famous Renaissance painter; he wrote mathematical treatises as well):

> A pound of silk costs 5 *libre* and 3 *soldi*, how much are 8 ounces? You have to multiply 8 ounces by 5 *libre* and 3 *soldi*. This makes 41 *libre* and 4 *soldi*. Dividing by one pound is no good, this works by converting to ounces, which are 12 per pound. Now divide 41 *libre* [and 4 *soldi*] by 12…. (della Francesca 1970, 43)

According to the rule of three, we should take the known price (5 *libre* and 3 *soldi*), multiply by 8 ounces, and divide by 1 pound. But for this to work, we need homogeneity of units. So we need to convert 1 pound to 12 ounces.

There seems to be nothing remarkable here. The only thing I'd like to point out is that in specific contexts of calculation, such as this one, some numbers may have to be replaced by others (here 1 must be replaced by 12). The symbol "1" does not necessarily represent its face value in the context of this calculation. In this specific context, it actually stands for 12. So in a very unremarkable manner, sometimes numbers (independently of unknowns and variables) stand for values other than their face values.

The next quotation is from Maestro Benedetto's *Tractato d'Abbacho* (Arrighi 1974, erroneously attributed to Pier Maria Calandri). This is the same Benedetto from the problem that opened this chapter. Since he wrote on practical mathematics in a commercial context, it's not surprising to find in his treatise a chapter devoted to currency conversion.

> The *fiorino*, to this money there is no fixed value, because sometimes it goes up by a few *scudi* and sometimes it goes down in price. At present it is worth around 6 *libre*, 3 *soldi* and 4 *denari*. There is of course an imaginary value that's called golden *scudo*, which is always stable, so that the *fiorino* is worth 20 golden *scudi* and the golden *scudo* 12 golden *denari*. (Arrighi 1974, 34)

The fact that money has a fluid exchange rate is taken for granted nowadays. But to note this fact in real time requires a relatively evolved commercial society that works with several kinds of coin, and is sensitive to the economical volatility of other economies and to the relative scarcity of metals. In this kind of situation, one may introduce imaginary monetary units in order to be able to draw standard calculations regardless of these volatile exchange rates. These imaginary units do not correspond to any actual coin (Einaudi 2006).

We see here a very trivial manifestation of a sign that changes its value, and a practice of inventing ideal magnitudes that serve as aids for calculation. Nothing here is very striking, but this system of variable values and of terms without worldly reference is required for practical mathematical economy even before we introduce unknowns and variables.

The last quotation is slightly trickier, at least for a contemporary reader. It reads:

> If 4 were the half of 12, what would be the ⅓ of 15? (della Francesca 1970, 48)

For us, this makes no sense—4 simply isn't half of 12. But the context can reveal what's going on. This, like the previous quotation from Piero, is an example of the rule of three. It's about conversion. It is an abstraction from questions like "if 4 pounds of silk were worth half of 12 soldi … ," or "if 4 *fiorini* were worth today half of what 12 *fiorini* were worth a year ago … ." This kind of abstract, seemingly counterfactual, conversion problem was typical of Iberian treatises, and was found in Italy as well (Høyrup 2012). Once again, we see here that a number sign need not stand for its face value.

So *even before we come up with letter signs for unknowns and variables, we already have economic and arithmetical practices where mathematical signs stand for values other than their face values, where they stand for variable values, and where they lack a real object as their reference.* Once we superimpose these practices on letters and ligatures that represent economic values, such as those shown earlier, we have the symbolic practice that will give rise to unknowns and variables (a further analysis of this process is available in Wagner 2010a).

First Intermediary Reflection

How does the story told so far fare with respect to the philosophies of mathematics surveyed in the previous chapter? We can definitely find support in the earlier narrative for the empiricist philosophical understanding of mathematics as an abstraction derived from worldly observations. However, while we usually think of mathematics as abstraction from natural science observations (especially physics), here the abstraction emerges from the realm of a social science: economy.

While observations emerging from natural sciences seem to reflect a given natural order, economic order is usually considered as contingently formed by humans. It is therefore not clear how to place these observations and their abstractions in a dichotomy between an empiricist natural order and the free conception of humans.

If we try to examine this narrative with respect to realist views, difficulties occur as well. On the one hand, if mathematical values derive from economic practice, where signs have variable values, they are nothing like the stable and ideal Platonic entities. However, we saw that sometimes ideal stability of value has to be imported into economic practice (in our case, the golden *scudo*). The appeal to idealizing imagination in this context could be interpreted as an appeal to an ideal reality, detached from the fluid world of phenomena. (We observed this attitude in the context of practitioner's remarks on the Black-Scholes formula as well.) So the mathematical world we're describing has to do with both Platonist and empiricist conceptions.

If we try to think in Kantian terms instead, we might consider the main tool of abbacist calculations, the rule of three, as an a priori rule. Indeed, it cannot be empirical, because empirically, the price of a product is often *not* directly proportional to its quantity—this is the very logic of bulk purchase discount. So the rule of three is not an inductive generalization, but a normative standard that gives regulative sense to the very notion of pricing a different amount of the same commodity. (This claim is elaborated in a different manner in the analysis of the concept of "value" in Heeffer 2011.) While it is a priori, the rule of three is not included in the concept of pricing, and is therefore synthetic.

The rule of three can also be considered synthetic a priori in the sense that it assumes an intuition of temporality inherent in the concept of value; indeed, economic values require temporality to fluctuate. However, traditional synthetic a priori interpretations of mathematics draw a sharp boundary between fixed numbers and fluid variables, which is not available in the abbacist context ("if 4 were half of 12 … "). The mathematical world of the abbacists, being a reflective standardization of economical practice, lives with monstrous hybrids between fixity and mobility that the Kantian scheme would not tolerate.

Abbacist mathematics is indeed world-making, and many authors since Marx have argued for connections between the economic conceptions of value in early capitalism and the quantified, computational, and alienated world that we live in today. But this is a contingent neo-Kantian form of world making, far removed from Kant's transcendentally deducible synthetic a priori. Kantianism, like empiricism, nominalism, and Platonism, are all reflected in abbacus mathematics, but only as partial aspects of a more complex phenomenon.

The Arithmetic of Debited Values

The problem that we opened with, about the five men finding a purse, is, as noted earlier, presented already in Fibonacci's 1202 *Liber Abaci*. One notable feature of this question is that it is among the earliest European examples of problems with negative solutions (Sesiano 1985, 118).

Fibonacci figured that the money of the first person (q_1) together with the purse (b), being $2\frac{1}{2}$ times as much as all the others, must constitute $\frac{5}{7}$ of the total amount of money between the purse and all five people ($q_1 + \cdots + q_5 + b$). Putting together the analogous information from the declarations of the other people, one obtains identities of the (anachronistic) form $q_1 + b = \frac{5}{7}(q_1 + \cdots + q_5 + b)$; $q_2 + b = \frac{3}{13}(q_1 + \cdots + q_5 + b),\ldots, q_5 + b = \frac{6}{43}(q_1 + \cdots + q_5 + b)$. Summing up these identities, Fibonacci could find the ratio between the money in the purse (b) and the total between all the people and the purse ($q_1 + \cdots + q_5 + b$). He then posited the value 1,455,636 for the total amount of money—a number divisible by all the relevant fractions, so as to guarantee a solution with whole numbers—and proceeded with the calculation (note that the problem has six unknowns but only five equations, so it is underdetermined, and an arbitrary determination of one of the values is permitted). The purse then turns out to contain 1,088,894; as for the money of the first person:

> [T]here will be 1,455,636 that is the amount of the purse and the denari [money] of the five men, and because the first has $\frac{5}{7}$ of the entire sum, you take $\frac{5}{7}$ of the 1,455,636 that is 1,039,740, and this many the first has with the purse. But because above more is found in the purse than is had between the purse and the first man, this posed problem will not be solvable unless the first man has a debit, namely that which is the difference between his amount plus the purse and the entire amount of the purse, namely that which is the 1,088,894 minus 1,039,749, that is 49,154. (Sigler 2002, 321–22)

Since the money of the first person and the purse turned out to be less than the money in the purse alone, the first person has to have a debited value, interpretable as debt. But this concern arises only when the final result is obtained, and does not intervene in drawing the solution.

Benedetto's version is not a translation of Fibonacci's (although he was aware of Fibonacci's precedence, and even quoted his result, which is different from Benedetto's, because of the previously noted underdetermination). The general idea, however, is similar, and leads Benedetto to derive the following relation between the *quantità* of the money of the first man, and the amount of money in the *borsa*.

> ... 1 *quantità* $\frac{344078}{1819545}$ and 1 *borsa* $\frac{1071896}{1819545}$ are equal to one *borsa* $\frac{2}{5}$. If next you remove from both sides 1 *borsa* $\frac{1071896}{1819545}$, you have that $\frac{7622258}{1819545}$ *quantità* are equal to $\frac{344078}{1819545}$ *debito de borsa*. In order not to have fractions, multiply each

side by 1,819,545. We'll have that 7,622,258 *quantità* is equal to 344,078 *borse debito*. Therefore, the 344,078 *borse debito* are worth as much as the 7,622,258 *quantità mobile*. Here it is clear that the *quantità* will have a *debito* value. In order to have whole numbers you say that the first [person] will be worth *debito* 344,078, and the *borsa* will be worth *mobile* 7,622,258. (Siena, Biblioteca Comunale degli Intronati, L.IV.21, fol. 264r)

Here the *debito* (debited) and *mobile* (cash) values are not just qualifications pinned onto the solution at the end, after all the algebraic reasoning is done. They enter into the equations themselves. Moreover, the money of the other persons is calculated in terms of the *quantità*, so the debit is not only an output, but also an input for further calculations. In the 250 years separating Fibonacci and Benedetto, *debito* has turned from a description of the result of a calculation into an element inherent in mathematical practice. How can we account for this change?

I believe that there are two complementary processes here. On the one hand, numbers are determined by their *nature* or *species*. Numbers may be wholes, fractions (halves, thirds, quarters), monetary units (such as *fiorini*), measures of length or area, temporal units, algebraic unknowns, and so on. (Often enough, even the root of a number is considered as that number having the nature of a root, rather than as an operation applied to the number.) The nature or species of a number affects its value: 4 yards of silk, 4 halves, and 4 unknowns may obviously have different values. The rules governing a mathematical sign and its value depend on its nature, so each species of quantity has its own rules, and these rules are often presented separately. In the preceding problem, debit can be considered as yet another nature that a number can have, affecting its value and how we operate on it.

On the other hand, the nature of a number may be converted. (This is one reason why, as shown in the first part of this chapter, the value of a number is not necessarily its face value.) For example, when making calculations, everything has to be brought to the same nature so as to provide a single bottom line. You can't put together different currencies or fractions unless you convert to a common currency or a common denominator. When multiplying a whole and a root (for example, $2 \times \sqrt{3}$), the calculators were instructed to convert both to the nature of a root ($\sqrt{4} \times \sqrt{3} = \sqrt{12}$). This practice leads to some strange procedures, such as bringing fractions to a common denominator before multiplying them—an apparent waste of effort from a contemporary point of view.

It is this practice of conversion that allowed integrating debits into the common scheme of mathematical values as subtracted values. I argued (Wagner 2010c) that these two seemingly contradictory processes— (ideal) division of quantities according to distinct species versus (empirical-nominalist) fluid convertibility of values—were constitutive of the evolution of Renaissance algebra. But how did negative numbers even become a species of quantity?

Albrecht Heeffer (2008) offers an interesting explanation. We begin with one species of quantity in abbacus mathematics: the binomial— namely, the sum or difference of a number and a root. Belonging to a distinct species, binomials required their own rules, including multiplication rules of terms such as $(a - \sqrt{b})(c - \sqrt{d})$. The discussion of the multiplication of the two subtracted roots (referred to as *meno*, literally "less") in a product of such binomials was extracted from this context, giving rise to independent rules for *meno* numbers, including the famous "minus times minus is plus." Extracted from binomials, *meno* became a new species of number, carrying with it the rules derived from the rules governing binomials.

Heeffer's argument can be further substantiated by an example that shows this process in "real time." It is taken from Maestro Dardi's mid-fourteenth-century algebra (Dardi 2001; see also Van Egmond 1983). The example I'm about to quote follows Dardi's list of 194 cases of equations, all reducible (in contemporary terms) to linear or quadratic equations. Since different species of numbers (wholes, roots, *meno*) were treated as distinct, equations whose coefficients belonged to different species were to be treated separately, which led to Dardi's proliferation of cases.

At the end of this exhausting (but not nearly exhaustive) list of cases and examples, come two remarks, which are not numbered as distinct cases (their position might suggest that they represent a later addition). The first remark considers equations of the form "unknown and number equal root", such as $5C + 6 = \sqrt{90}$ (C, short for *cosa*—literally "thing"—is Dardi's representation of the unknown quantity). To solve this, one subtracts 6 from $\sqrt{90}$, and divides the resulting binomial $\sqrt{90} - 6$ by 5, yielding $\sqrt{3\frac{3}{5}} - 1\frac{1}{5}$.

Next, without any fanfare or ado, we have the following ground-breaking paragraph, possibly the first in Europe to thematize a negative or debited solution independently of an explicit economic context:

> Know that when you get C and numbers on one side equal to nothing (as some equations can be in many of the cases that we've discussed, many of which involve a difference or other things), you have to subtract the numbers from both sides, and you'll get C equal nothing *meno* number, in which case you have to divide the number which is *meno* by the quantity of the C. Whatever comes from this is the C *debita*. Suppose that you get 5C and 15 n[umbers] equal to nothing. Divide the 15 by 5, there comes 3, and such will be *debita* the unknown [*cosa*]. (Dardi 2001, 297; my translation depends on the Arizona manuscript described in Hughes 1987, which is being edited by Warren Van Egmond).

The equation $5C + 15 = 0$ is solved by first reorganizing it as $5C = 0 - 15$, and then dividing the subtracted term by 5. The points made in the quoted paragraph are the following. First, the case "unknown + number = nothing" is not an artificial addition, but something that may emerge from manipulating equations by means that already belong to our scientific arsenal (indeed, something like this happened in Benedetto's solution of the purse problem). Second, the method of solution is presented as perfectly analogous to the preceding case (the preceding binomial "root less number" is replaced here by the binomial "nothing less number"). In the final step, the subtracted number in the binomial is translated from a *meno* (less or subtracted) value into a debited value. We see here a direct and explicit link between binomials and the emergence of what will become, within a few centuries, fully fledged negative numbers.

When roots of negative numbers later emerge in Raffael Bombelli's mid-sixteenth-century solution of cubic equations, a similar process occurs. Specifically, the dal Ferro–Tartaglia formula for solving the equation $x^3 = 15x + 4$ states that the solution should be $\sqrt[3]{2 + \sqrt{4 - 125}} + \sqrt[3]{2 - \sqrt{4 - 125}}$. This was rejected as senseless by most sixteenth-century mathematicians. Bombelli's leap forward was to rewrite the difference $4 - 125$ as the binomial $0 - 121$, and then use his version of more or less known considerations for extracting cubic roots of binomials (deriving equations for determining the integer and root part of the binomials, and looking for integer solutions by trial and error). The sum of cubic roots was thus reduced to $(2 + \sqrt{0 - 1}) + (2 - \sqrt{0 - 1})$. Canceling the roots, one gets the result 4. Substituting this 4 into the original equation shows that it forms a correct solution.

La Nave and Mazur (2002) argue that Bombelli's roots of negative numbers are introduced as a new species of number. Just as we have binomials (including binomials of the kind "nothing less number"), just as we have roots, and just as we have roots of binomials, we may introduce into our list of species roots of "nothing less numbers" binomials. We can then formulate the rules of these binomials by analogy to other binomials, and use them to solve equations.

Second Intermediary Reflection

We saw earlier that the values of abbacist numbers and unknowns are not necessarily their face values. Even if they appear to be fixed numbers, their values may fluctuate. They come from an economic reality, but are not always empirically referential (for example, the imaginary golden *scudo*, or solutions in terms of irrational roots to pricing questions). This makes it difficult to fit them into Platonist/empiricist or realist/nominalist dichotomies. The rule of three, which governs the conversion of these values, can be viewed as some sort of a synthetic a priori foundation governing quantities. But given its contingent applicability, it can't be viewed as a necessary transcendental truth.

The discussion of negative numbers in the previous section adds to these tensions. We see that the formation of negative numbers operates along three axes. The first is an empiricist-economical axis: the notion of debt and its integration into calculations. The second axis is Platonist: the division of quantities into ideal species, each of which has its own ontological status and rules, including binomials, *meno* terms, and their roots. The third axis is formalist: the derivation of mathematical rules from axiom-like hypotheses about species of quantities.

To make the latter axis more salient, consider Dardi's proof that the product of *meno* terms is additive. He considers the product $(10 - 2) \times (10 - 2)$. On the one hand, this should be identical to $8 \times 8 = 64$. On the other hand, implicitly assuming distributivity, Dardi breaks this product into 10×10, two terms of the form 10 times *meno* 2, and a product of a *meno* 2 with itself. The first three terms yield $100 - 20 - 20 = 60$. The only way to reach the expected result of 64 is by stipulating that the product of a *meno* 2 with itself is an added 4. In other words, minus times minus is plus. Similar reasoning applies to Bombelli's work with roots of negative numbers.

In this case, the rules governing a species of quantity turn out to be derived formally from implicit principles such as distributivity, rather than from a direct understanding of the underlying species. But the crucial point is that this formalist argument does not stand alone. It is superposed on an ontology that assumes a variety of species in the genus of quantity, including the *meno* species, and on an economic context of translating species of quantity into economic values (such as debts).

It is sometimes argued that the history of mathematics is a history of doing away with unnecessary divisions into species of quantities (namely, arithmetizing all the way down), and of abstracting beyond any empirical context. But this narrative ignores the productivity of these rejected practices. For instance, the insistence on considering numbers, roots, and binomials as distinct entities actually helped understand the solvability conditions of equations and find new solutions. Indeed, it was by refusing to reduce numbers, roots and binomials (a, $\sqrt{b}$, $a + \sqrt{b}$, $a - \sqrt{b}$) to a homogeneous number concept, that some kinds of equations with positive integer coefficients could be proved not to have solutions consisting of square roots or binomials (one would substitute a certain species into the equation, and show that the result could not be a positive integer). Eventually, this exploration of the relations between the natures of coefficients of equations and the nature of the solutions of those equations probably helped guess the expected species of the solution and actually solve cubic equation (for details, see Wagner 2010a, §4).

So attempting to reduce mathematics to a single foundation (formal, empiricist, arithmetical, and so on) does not do justice to mathematical reasoning. Attempting to reconstruct a normative ideal for mathematics by promoting one foundational axis at the expense of others is a problematic venture. This kind of reconstruction may help convince us that mathematics is more likely to be consistent, but drives us further away from an understanding of how mathematics works.

It is precisely the superposition of different aspects of mathematical entities (in this section, empiricist, Platonist, and formalist) that makes mathematical entities rich, useful, and robust. They don't depend on just one foundation, but acquire their support from several anchors. If one foundation falls apart—if the Platonist division of species of quantity is shaken, or if we fail to find an empirical manifestation of some mathematical entity or maneuver, or if we need to readjust formal rules because they end up contradicting each other—the other pillars that support mathematical reasoning allow it to survive and evolve.

False and Sophistic Entities

Negative numbers and their roots were subject to an ambiguous attitude in Italian algebra. As algebra evolved and solutions of cubic and quartic equations were published by Cardano and Tartaglia, negative numbers became more widespread and naturalized than in earlier abbacus treatises (and so I allow myself to use the anachronistic term "negative" rather the indigenous and ambiguous *meno*). Cardano and Bombelli, for instance, did resort to negative numbers in their work, but not without reservations.

Both Cardano and Bombelli acknowledged negative numbers as interim results in the process of solving more complex problems. Occasionally, they allowed negative numbers to stand even as final results. But at other times these same authors also tried to circumvent and avoid negative numbers, calling them false and sophistic. The result was an inconsistent mix of affirmations and denials of negative numbers (Wagner 2010c, §5).

Cardano, the more philosophically inclined of the two, had a significant change of heart concerning negative numbers as he approached the end of his life. In the famous *Ars Magna*, he stated, like his predecessors, that a minus times a minus makes a plus. But later he repudiated this claim, and argued that the product of minus terms has to be a minus (Tanner 1980a). This view was not Cardano's alone. It has traces in authors such as Marco Aurel and Piero della Francesca as well (Wagner 2010c, §5; this theme was later carried on by Thomas Harriot, see Tanner 1980b).

But how does this fare with respect to the proof quoted from Dardi earlier—a proof with a sound geometric interpretation that Cardano could not ignore? Indeed, Cardano did not deny that $(10 - 2) \times (10 - 2) = 100 - 20 - 20 + 4$. He denied that this calculation had anything to do with the product of negative numbers. In other words, he denied the distributive identity $(a - b) \times (a - b) = aa + a(-b) + (-b)a + (-b)(-b)$. This may strike us as impossible, but if we do not take for granted the distributive law and identities such as and $a - b = -(b - a)$ and $a - b = a + (-b)$, a notion of negative number as distinct from subtracted number can be rendered consistent. Such an endeavor was indeed elaborated in modern terms by Martínez (2006, 133).

Cardano's argument was ontological. Positive numbers were supposedly real, and negatives were "alien" to reality. The product of things alien to reality could not be real, and so the product of negatives had to be negative. If higher algebra were significantly employed in theology or ontology, this argument might have had some impact. But this was not the case, and so distributivity prevailed. Subtracted numbers were superposed on negative ones, and the algebra that could have developed from Cardano's arguments was nipped in the bud.

Bombelli's approach was more instrumental. It is best exemplified in his treatment of the equation $x^3 + 165 = 9x^2 + 9x$. In order to solve this equation, Bombelli transformed it, and obtained the equation $y^3 = 36y - 84$, where $y = x - 3$. (Today, we'd call it a change of variable; this was not Bombelli's terminology or notation, but for this context it will do.) In Bombelli's terms, the latter equation is "either impossible or the derivation of the equation was badly done" (Wagner 2010c, 504). In our terms, this equation simply has no positive solutions. Nevertheless, Bombelli goes on to state that "the value of the [y] being found (if it could be), if we would add 3 … the sum would be the value of the [x]" (Wagner 2010c, 504).

This statement hardly makes sense. What does it mean to add 3 to a value that does not exist? There are two explanations here. First, Bombelli was perfectly aware of negative solutions, and was proficient in handling them. He went through examples where a change of variable changes a negative solution into a positive one. This, however, is not the case here—adding 3 to the negative solution of the new equation (for y) would still produce a negative solution of the original equation (for x). This was therefore useless for Bombelli.

The alternative explanation is that Bombelli dealt here not with a single isolated equation, but with an example for a general procedure. If the coefficients of the problem had been different, the procedure might work. The "165," "9," and "3" here did not stand just for their face values; they stood for all possible coefficients, including those that would allow the former method of solution to produce a positive solution.

This is, again, the principle of fluidity of mathematical signs. As mathematical signs change their values according to context, they may change their own nature or species, or the nature or species of related entities, such as solutions of equations. A negative solution may become positive, if the terms of the problem are changed. Therefore, a "false" or "sophistic" solution cannot be discarded out of hand. Nonexistent entities can be treated arithmetically, because a change of context may turn them into something real (as in this context), or because they can facilitate calculation and can eventually be converted into a real result (as in the case of the imaginary golden *scudo*). Now if this kind of logic is applied to roots of negative numbers, it is no longer very surprising that Bombelli dared to go through them on his way to real solutions of cubic equations.

But this story has to be refined. In the solution of the cubic equation $x^3 = 15x + 4$ considered earlier, which went through roots of negative numbers, the sophistic roots of negative numbers eventually cancelled out and yielded the real solution 4. In other examples, the cubic roots could not be transformed by Bombelli's techniques into a real result, and the only available representation of the solution involved roots of negative numbers.

Bombelli wasn't satisfied with this situation. To validate it, he sought a geometric representation. Indeed, he elaborated an entire system for translating algebra into geometry (Wagner 2010b; and the section on Bombelli in chapter 6). Among other things, he showed how to construct the solution of a cubic equation using a right-angled ruler. This construction of a geometric magnitude that represented the solution of the equation did not provide representations for the roots of negative numbers involved in the complex arithmetic representation of the real solution, but for Bombelli this form of representation was enough to lift the veil of "sophistry" that hung over his work.

Final Reflection and Conclusion

This part of the story is mostly about handling monsters. As the negative "monsters" began to emerge in a prosperous mathematical domain, abbacists simply lived with them, handling them ad hoc, while being aware of their dangers.

As their role became more central, we see more serious attempts to bar or contain them. Cardano's later attempt was an attempt at monster barring. First, negative numbers were confined in the realm of the "alien." Second, if we accept that the product of negatives is negative, then so are the roots of negative numbers, and the problem of negative roots of positive numbers together with the whole issue of complex numbers is circumvented. (This aspect of Cardano's reasoning is highlighted by Tanner 1980a.)

Bombelli's strategy, on the other hand, was first to live with monsters: he thought that they were sophistic, but useful in obtaining correct results. Later he attempted to translate them to the well-established field of geometry by geometric constructions. As is typical of classical geometry, his approach has a distinct constructive streak: mathematical elements are admissible because they can be constructed from a small pool of elements and basic construction moves. However, for Bombelli's constructions, ruler and compass were not enough; he had to adjust the notion of construction by including a right-angled ruler. Constructivity as criterion for monster barring or taming was a rather vague and flexible notion; this was true of early modernity (Bombelli, Descartes) as it was true of the turn of the twentieth century (Baire, Borel, Lebesgue).

Cardano's attitude turned out to be on the losing side, but I argue that the reasons for the defeat are contingent. As I said, had there been a serious attempt to develop a mathematical theological ontology in Cardano's time, his approach might have yielded an algebraic structure different from the one that emerged from Bombelli's work. (If the notion of doing ontology by mathematics strikes you as absurd, then check out the recent work of Alain Badiou.)

The two algebras—the algebra of subtracted numbers and the algebra of alien negative numbers—could have lived side by side, networking with various relevant domains of knowledge; or perhaps one of them—either one—could have suppressed and absorbed the other into its own terms. But we know that at the time the religious authorities (or at least the powerful Jesuits) preferred a mathematics that affirmed the old order over the introduction of new, suspect entities. The social coalition of successful mathematicians involved the rising commerce-oriented classes, while older philosophical and theological authorities were fighting a slowly losing battle.

As for Bombelli, we should correct the impression that we're dealing here with the reduction of arithmetic monsters to geometry. In fact, Bombelli wrote that "these two sciences (that is Arithmetic and Geometry) have between them such accord that the former is the verification [*prova*] of the latter and the latter is the demonstration [*dimonstration*] of the former" (Wagner 2010b, 234–35). The point is not to reduce one to the other, but to combine the two.

As I argued (Wagner 2010b; see also the section on Bombelli in chapter 6, later), this unification was not entirely conservative; it had to adjust the content of both geometry and algebra. But it is precisely because of this transformative unification that both geometry and algebra could be made more robust. Where arithmetic was edgy, one could turn to geometry for proof; where geometry was shaky, one could turn to arithmetic for verification. The project here is not a project of foundation, but a project of thickening the network of relations that constructs the relevant mathematical objects. In fact, throughout the history of the arithmetization of mathematics, we can find mathematicians lamenting the loss of genuinely geometric points of view (for example, Poncelet, Poincaré, the early twentieth-century Italian school of algebraic geometry). The technical victory of arithmetic at the front of foundations, however, does not reflect the actual practice of mathematicians who depend heavily on geometric intuitions.

A final observation concerns the issue of authority. We saw that a coalition of mathematics and old-school theology did not emerge. However, even the pragmatic Bombelli attempted to create a coalition with the classical authorities. In the fifteen or twenty years between his manuscript and print edition, Bombelli became acquainted with Diophantus's work. His reaction was to change his terminology and practice problems to come closer to Diophantus's more abstract arithmetic presentation. Like the attempt to unify geometry and algebra, the attempt to unify abbacist algebra with the Diophantine style was meant to strengthen the emerging science and confer classical Greek authority (which has been gaining respect throughout the Renaissance) on new ideas.

The purpose of this chapter was to follow abbacus and Renaissance algebra as a case study for what I called at the end of the previous chapter the "yes, please!" philosophy of mathematics. Rather than decide between philosophical approaches and the divisions that underlie them, I wanted to show how competing philosophical approaches find an intertwining expression in mathematical practice. I prefer to view different philosophical approaches as descriptions of aspects of mathematical practice, rather than choose from among them a foundational or monopolizing explanation of what mathematics is. Platonism or empiricism; nominalist, formalist, and constructivist approaches; synthetic a priori views; attempts to bar or tame monsters; conflicts or coalitions with other intellectual authorities—these are all parts of how mathematics works in very concrete and down to earth ways.

I am aware, however, that for many, such an eclectic approach to the philosophy of mathematics is not enough. The next chapter will therefore attempt to formulate a philosophical approach to mathematics that can serve as an integrative framework for the insights of the various philosophies of mathematics.

■ ■ ■ ■ ■ ■ ■ ■

CHAPTER 3

A Constraints-Based Philosophy of Mathematical Practice

THIS CHAPTER WILL TAKE a detour around the host of problems discussed in chapter 1 in order to deal with them from a different perspective. Instead of asking foundational questions about the grounding of mathematics, its freedom, its unique position, its monsters, or its source of authority, we will ask some questions about mathematical practice. How are mathematical statements used? How do people get to agree on them? How are they interpreted?

The resulting observations will be integrated into a constraints-based philosophy of mathematics: instead of debating the reality of mathematical entities, we will think of mathematics as a field of knowledge that negotiates various kinds of real constraints. This contingent array of constraints and the various ways of juggling them lead to the formation of various different mathematical cultures.

This perspective will then be brought back to explore the field of problems presented in the first chapter together with the issues of reality and truth. But our detour will shift the focus and span of these problems, and suggest a different outlook on managing them.

Dismotivation

The point is that the proposition $25 \times 25 = 625$ may be true in two senses....

First, when used as a prediction of what something will weigh—in this case it may be true or false, and is an experiential proposition. I will call it wrong if the object in question is not found to weigh 625 grams when put in the balance.

In another sense, the proposition is correct if calculation shows this—if it can be proved—if multiplication of 25 by 25 gives 625 according to certain rules.

It may be correct in one way and incorrect in the other, and vice versa.

It is of course in the second way that we ordinarily use the statement that $25 \times 25 = 625$. We make its correctness or incorrectness independent of experience. In one sense it is independent of experience, in one sense not.

Independent of experience because nothing which happens will ever make us call it false or give it up. Dependent on experience because you wouldn't use this calculation if things were different. The proof of it is only called a proof because it gives results which are useful in experience. (Wittgenstein 1976, 40–41)

At a first glance, what we have in this passage is a version of the logical positivist position: a mathematical statement can be used synthetically, as an experimental statement, or analytically, as a statement that can be derived from concepts and rules. But Wittgenstein adds to the story a temporal-causal element: first the proposition or rules of multiplication are adopted because they provide a successful empirical description of practical counting, weighing and measuring. Only subsequently, because of its a posteriori success, does it become a rule that no experiment can refute.

I refer to this process as "dismotivation," indicating the gradual loss of a mathematical statement's empirical motivations and grounding. We have already seen it at work in the previous chapter: subtracted numbers, which were first only subtracted from larger numbers, and later empirically related to debts, were isolated and carried out of these contexts, and turned into new independent entities: negative numbers. This process was so successful that things go the other way round in contemporary commutative group theory: when using additive notation, $-a$ is defined as the reciprocal of a, and one doesn't usually even define subtraction. If it is defined at all, $a - b$ is defined as $a + (-b)$, so negative numbers turn out to precede subtraction.

To better understand the process of dismotivation, consider Wittgenstein's following example:

All the calculi in mathematics have been invented to suit experience and then made independent of experience.

Suppose we observed that all stars move in circles. Then "All stars move in circles" is an experiential proposition, a proposition of physics. Suppose we later find out they are not quite circles. We might say then, "All stars move in circles with deviations" or "All stars move in circles with small deviations."

The simplest method of describing their paths might be to describe their deviations from the circles. Suppose I now say "All bodies move in circles with deviations," meaning [some arbitrary open curve] is a circle with deviation—now I am no longer making a statement of physics. It is now a proposition of geometry; I have made it independent of experience. I have laid down a proposition which provides a form of representation, a method of description....

It is the same with $25 \times 25 = 625$. It was first introduced because of experience. But now we have made it independent of experience; it is a rule of expression for talking about our experiences. [If 25 objects weighing each 25 grams don't weigh together 625 grams, we'd] say, "The body must have got heavier" or "It deviates from the calculated weight." (Wittgenstein 1976, 43–44)

The point here is not only that we don't need experience to validate the mathematical rule. We actually use the rule as *a reference against which to measure experience.* As the star orbits example shows, this can still be done even if reality behaves very differently from our reference rule.

Indeed, this is demonstrated by the Black-Scholes vignette from the introduction: the result is empirically motivated by a very specific setting, and then becomes a reference point for other settings, where it may be very far from empirical observation. However, note that in the Black-Scholes case, it is hard to articulate a specific context where the result was validated by experience over a substantial span of time. It became dismotivated almost instantaneously.

There are many examples of elementary mathematics that deviate from the arithmetical standard: if we put together ice cubes, or droplets of water, or line segments on paper, the rules of addition and multiplication need not conform to the actual empirical count. The ice cubes may melt before we're done counting, the water droplets may pool together, several line segments may be drawn as continuous and form a single segment. These deviations could be seen as affirming the claim that arithmetic rules are a point of reference independent of experience.

But can we perhaps save the relation between mathematical statements and empirical experience, if we articulate more precisely the conditions under which the rules are empirically correct? Perhaps elementary arithmetic does apply empirically, provided that we restrict our empirical scope to a certain *way* of putting objects together, which precludes an interaction or dissolution of the counted elements as in the preceding examples?

I'm afraid, however, that restricting the scope of application by precluding the interaction of counted elements will not get things quite right. In an electric circuit, for example, if we put units of resistance together in a series, simple addition will give us the total resistance. If we put them in parallel, it won't. In both cases, there obviously is an interaction of resistances in the electric circuits, violating the "no interaction" criterion. But this violation won't prevent us from saying that addition applies to serial combination of units of resistance. So restricting the empirical applicability of addition to non interacting entities is too restrictive.

Here's another example: one unit of luminescence and one unit of luminescence put together will measure two units of luminescence only if the light sources are allowed to interact. Indeed, if we keep the light from the two sources from interacting by an opaque barrier that allows each point in a room to be exposed to only one light source, the measured luminescence anywhere in the room will not increase compared to its level with only one source of lighting. In this example, we see that addition would apply only if we violate the no-interaction restriction.

In the earlier examples, the no-interaction rule proved too restrictive—it prevented us from applying addition where we would normally like to apply it. But it is also not restrictive enough, even in the context of a simple count of discrete, inert objects. Indeed, if we are dealing with very large numbers, we are not likely to reach the expected arithmetic results by counting (we'd say there are so many objects, we're bound to have counting errors). When numbers get really large, entropy considerations make even mechanized counting impossible according to physical principle, not only in practice (Rotman 2000, ch. 3). But we still adhere to elementary arithmetic, even when it bears no relation to counting objects.

Of course, we could attempt to fine-tune the statement of conditions under which arithmetic calculations should reflect the aggregation of empirical objects. But I doubt that such a venture would be successful—it will most likely involve some vagueness of interpretation that would limit its claimed success. Even more important is the fact that such a venture is not considered as terribly important. Instead of drafting rules that determine in advance which ways of aggregating objects actually fit arithmetic calculations, we more or less just know that addition applies empirically to those situations to which it turns out to empirically apply. Moreover, there are many branches of mathematics that have grown so distant from their applicable roots (abstract algebra, large cardinals) that the concern with conditions of empirical applicability is simply irrelevant.

Mathematical statements involve rules that may not be empirically descriptive, but instead set standards against which we describe empirical results. Moreover, even in its applicable parts, mathematics does not depend on the establishment of rules that tell us when we should expect mathematical statements to empirically apply. I believe that this form of dismotivation—not a break with empirical applicability, but a lack of interest in exhaustive rules for the general applicability of mathematical statements (as opposed to concrete rules in specific restricted circumstances)—is one of the deepest senses in which arithmetic is made independent from experience.

The Analytic A Posteriori

What is the status of mathematical descriptions turned independent of experience? Kant suggested that we consider them as synthetic a priori—that is, emerging from the structure of our senses, rather than from reason or empirical observation. I suggest that perhaps they're better subsumed under the title "analytic a posteriori": following from formal rules *and* dependent on worldly experience.

True, Kant does exclude the combination of analyticity and dependence on a posteriori empirical experience. He claimed that if a statement is true because it is analytic, that is, because the predicate is included in the concept ("a triangle has three sides"), then all I need to validate the statement is the concept and the predicate, regardless of experience.

Nevertheless, structural positions abhor vacuum, and there's a history of attempts to fill the empty box. Stephen Palmquist (1993, app. IV) has recorded several such attempts. One kind of statement that he considers as analytic a posteriori (following Saul Kripke, but rearranging his terms and definitions) is "naming." In order to name someone, we need to have an experience. But once they are named, the claim that they are so named is purely analytic—it is included in the name concept of the person so named.

I think that this usage of the analytic a posteriori does not quite work. When I call someone by her or his name, the challenge is to identify time and again that the person I address by that name is indeed the very person so named, and not just to figure out the analytic truth that the name applies to the person so named. The former challenge is a synthetic and a posteriori task.

So I'll go in a different direction (which may be closer to the articulation of the analytic a posteriori in Palmquist 1993, ch. IV, §3). When I use the combination of analytic and a posteriori, I refer to a statement that derives from synthetic a posteriori judgments but that has been rendered independent of experience. This reference concerns precisely the dismotivated mathematical statements discussed earlier.

One objection may be that what I am talking about is actually a synthetic a posteriori statement that has become analytic a priori, once a certain concept was formed. At no time, therefore, was the statement at once analytic and a posteriori. I reject this objection on two grounds.

The first has to do with our underdetermined use of mathematical statements. In some cases, it's very clear that I use some mathematical statement empirically, and in others, it's clear that I use the same statement analytically, as a purely formal statement. But in many cases, the distinction can't be made. When we learn that $25 \times 25 = 625$, we learn it as a statement with multiple justifications and uses. It is both the result of a calculation and a potential description. (I will elaborate on such ambiguities in the section on interpretation in this chapter.)

But even if we restrict our attention to the analytic meaning of the calculation, to the way it follows from definitions and rules of calculation (which may be reducible to rules of logic, if we accept the relevant subset of Russell's articulations), there's an a posteriori element involved. This a posteriori element is the actually performed calculation in an arithmetic or logical formalism by people or by machines. We derive the result analytically, and yet the derivation occurs in time and space.

Now, we are not used to thinking of the result of arithmetical operations as dependent on what we do in time and space. The result is considered as determined once and for all, we only gain access to it by calculation. Yet,

> Suppose that from now on, when we were told to multiply, we all of us constantly got different results. Then I suppose we should no longer call this calculation at all. The whole technique ... would lose the character of a calculation. We would then no longer in fact have a right or a wrong result.
>
> The whole thing is based on the fact that we don't all get different results.... [T]he agreement in getting this result is the justification for this technique. It is one of the agreements upon which our mathematical calculations are based. (Wittgenstein 1976, 102)

A necessary condition for considering a result as mathematical is that we tend to agree on the result. However, what is the nature of this agreement? Why is it that we usually get the same result? One reason for getting the same result may be that this result is true. But truth is never a sufficient guarantee for its actual revelation. There are many truths that cannot be derived by a straightforward calculation.

Wittgenstein considers this very point in a dialogue with Alan Turing (they discuss one-to-one correlations, rather than arithmetic calculations):

> WITTGENSTEIN: Suppose you had correlated cardinal numbers, and someone said, "Now correlate all the cardinals to all the squares." Would you know what to do? Has it already been decided what we must call a one-one correlation of the cardinal numbers to another class? Or is it a matter of saying, "This technique we might call correlating the cardinals to the even numbers"?

TURING: The order points in a certain direction, but leaves you a certain margin.

WITTGENSTEIN: Yes, but is it a mathematical margin or a psychological and practical margin? That is, would one say, "Oh no, no one would call this one-one correlation"?

TURING: The latter.

WITTGENSTEIN: Yes.—It is not a mathematical margin. (Wittgenstein 1976, 168)

According to this position, mathematical techniques are formed in such a way that natural and practical considerations (the more or less common capacity of humans to be *trained* to follow certain kinds of orders in similar ways) guarantee more or less consensual results. Derivation techniques (including calculations or proofs) that fail to produce such consensus are not considered mathematical.

It is in this very sense that the purely analytic derivations are an a posteriori feat. Human conceptual analysis is conditioned on an a posteriori, natural, and social human capacity to obtain some kinds of consensus by following some kinds of rules. It is in this sense that mathematics is deeply analytic and a posteriori at one and the same time.

In thinking of mathematical statements as synthetic a priori, we consider them to be reflections of the pre-empirical formal intuition through which we view the world, and that leads us beyond logical and conceptual analysis. In thinking of mathematical statements as analytic a posteriori, on the other hand, we allow that they may indeed become so thoroughly dismotivated that they turn into logical facts, but we further claim that deriving logical facts through calculations or proofs depends on a posteriori common features of situated human practice and learning capacity, rather than on an a priori way of seeing things.

The analysis here places mathematical statements on a continuum between synthetic and analytic a posteriori. This is the continuum that ranges from empirical observations to following rules that are designed to be followed successfully by people (or that can have people trained to follow them successfully). Genuine a priori, pre-empirical knowledge cannot depend on such contingencies of practice. Genuinely a priori knowledge has to be obtained by pure thought, whatever that may mean, not by spatio-temporal performances of derivation procedures that tend to produce common results when followed by well trained practitioners or well constructed machines. Mathematics as a whole is much too dependent on empirically grounded semiotic activity to be confined to the realm of the a priori.

Consensus

One of the most impressive things about mathematics is that mathematicians tend to agree with each other over the validity of mathematical arguments. While they may strongly disagree over the importance, ingenuity, or originality of mathematical statements, their consensus over validity is not only infinitely greater than what we're used to in the humanities and social sciences, it also tends to be considerably greater than the level of agreement obtained in the natural sciences.

This agreement is, indeed, not absolute. Mathematicians do sometimes disagree about the validity of proofs. Long-standing proofs may be found to be invalid after many years. Some proofs are so long and complex that mathematicians hesitate to rule whether they are valid or not. Still, in contemporary mathematics, long-standing disputes and errors concerning the face value validity of arguments are the exception, not the rule.

While the vast majority of mathematical everyday disputes are resolved by some sort of semi-formal shorthand, relying on subject specific tool boxes or inference packages (Netz 1999; Azzouni 2005) combined with iconic representations, analogies, intuition, authority, and experience, this alone is not what allows mathematics to be much more consensual than other sciences. The consensus among mathematicians about the validity of proofs has a lot to do with formalization. By formalization I do not mean the translation of an entire proof into a strictly formal language (which is almost never done, and is in fact impossible for finite humans to achieve in the context of typical research mathematics). By formalization I mean a gradual process of piecemeal approximation of formality that is conducted only as far as required to resolve a given dispute.

The mathematical arbitration of validity disputes by means of partial formalization is more accessible and conclusive than the comparable means available to other sciences, and leads to a higher degree of consensus. This has to do with the following facts: that valid mathematical arguments are those that can be split up to subarguments, each of which (but usually not all at once) can be formalized by trained professionals with reasonable effort; that the process of verifying the validity of formalized proofs is explicitly designed to rely only on rules that people can be trained to follow with similar results; and that a process of partial formalization serves as the supreme arbitrator in cases of mathematical validity disputes (even if, like the Supreme Court, one rarely carries dispute all the way to that final instance).

This state of affairs seems to lend plausibility to the claim that mathematics represents a deeper truth, or stands closer to truth than other branches of knowledge. How else did it happen that mathematical arguments are, unlike those of other sciences, reducible to consensually verifiable formalizations? But the exceptional truth of mathematics may be put into question, if we consider the historical possibility that what guarantees consensus in mathematics is the active exclusion of arguments and concerns for which formal arbitration mechanisms are of no use. Mathematical consensus could arguably be the artifact of trimming mathematics along the dotted lines that allow consensus to be retained, rather than the expression of some essential mathematical trait.

Today, the final mechanism for arbitrating the validity of mathematical arguments is the rigid syntax of predicate calculus applied to formal axiomatic systems (even though, I must reiterate, this arbitrator is not appealed to very often). Historically, it was preceded by other semi-formal syntaxes. As noted earlier, scholarly Greek geometers had their own rigid syntactic code (Netz 1999), which, as Pasch and Hilbert showed in their analyses of the foundations of geometry, is not quite a formal system in the contemporary sense. Abbacus masters had calculation diagrams that generated unique solutions to various kinds of problems (even where today we consider these problems as having multiple solutions). Modernity introduced symbolic notations that set arguments into a uniform calculus, which in the nineteenth century were gradually appropriated for the purpose of creating a calculus of logic.

While the appeal to formal-syntactic validity arbitration mechanisms is not a twentieth-century invention, it would be wrong to consider it ahistorical. Indeed in many historical mathematical cultures, consensus was weaker than it is today. Moreover, the strict separation between the validity of an argument and its importance, style, and ingenuity is a rather novel feature of mathematics.

Indeed, the Greek mathematical scene was a polemic one, where a "mathematical text is a challenge: it attacks past mathematicians, and fully expects to be attacked, itself" (Netz 2007, 62). The sixteenth-century dispute between Tartaglia and Cardano/Ferrari is perhaps the most infamous debate in the history of mathematics. From today's perspective, the main issue is Tartaglia's precedence and publication rights, rather than mathematics. But if we look at the protagonists' own arguments, we find that they pool together the very correctness of proposed solutions, the efficiency of calculations, and the elegance of method in their respective attacks (Bortolotti 1933). This polemical style of mathematical disputation might explain why juridical linguistic structures were imported into mathematics at the beginning of modernity (Cifoletti 1992, 1995). Disputes that superposed mathematical validity, social cliques and generation gaps went well into the nineteenth century (Ehrhardt 2010, 2011; Wagner 2014, forthcoming). The Italian school of algebraic geometry was confronted with bitter and enduring disputes over issues of mathematical validity as late as the 1930s (Brigaglia and Ciliberto 2004). Ever since classical Greek geometry, European mathematics had a consensual mathematical core, but many open ended debates around it.

Predicate calculus, Hilbert's version of the axiomatic method, and the Zermelo-Fraenkel set theory made mathematics quantitatively much more consensual than it had ever been before. They allowed closing more debates than had ever been possible in Western mathematics. In fact, they were so successful that any question or argument that could not be submitted to this arbitration mechanism became a nonmathematical question, and was exported to other branches of knowledge. This is seen most clearly in the context of the early twentieth-century foundations "crisis." Those problems that could not be translated into one of the emerging formal systems were not consensually solved—they were expelled from mathematics and became philosophical.

The foundations crisis was not solved by providing a sound foundation for mathematics, but by exporting the problem of foundations. Any axiomatic foundation became acceptable, as long as it could be formalized. Mathematicians no longer had to agree on a set of axioms and formal systems; they could simply explore several prominent options (for example, with or without the axiom of choice, with or without large cardinals, Zermelo-Fraenkel versus Quine's *New Foundations*, and so on).

There are obviously many formal systems that are not considered worthy of investigation, and the criteria for selecting which formal systems should be studied are obviously not formalizable. But it is clear that whatever can't be formalized in one of the mathematically endorsed formal systems is no longer mathematical. Mathematics achieves consensus over the validity of proofs by rejecting all proof methods and articulations of questions that cannot undergo formalization and submit to formal mechanisms of arbitrating disputes. (That's why the liar paradox, the pop quiz paradox, and the sorites paradox are problems of *philosophical* logic rather than *mathematical* logic.)

Why have other sciences not achieved similar validity arbitration mechanisms? I believe that the answer lies in the process of dismotivation. In order to allow formal systems to be the final arbitrators of validity, whenever formal arbitration conflicts with empirical or other kinds of validity tests, the latter has to be demoted in favor of the former, and the scientific claims become dismotivated. Many sciences could not afford to become as dismotivated as mathematics. Their empirical reference had to be much more clearly defined for them to maintain their status as *empirical* sciences.

But some sciences did manage to come up with formal arbitration systems, even if it cost them the stability of their empirical reference. These branches of science were subsequently subsumed under mathematics. This was the fate of such fields as formal logic that has detached itself from actual reasoning, game theory that has been divorced from actual decision making, mathematical physics or economy that do not necessarily reject make-believe models just because they don't even pretend to approximate real-world phenomena, computational complexity that deals with asymptotics rather than finite scope results, and so on. Allowing myself to oversimplify, I might say that *the contemporary necessary and sufficient condition of being mathematical is precisely the combination of dismotivation with respect to empirical application and potential formalization serving as highest arbitrator in disputes over the validity of arguments.*

If this is so, then the twentieth-century increase in mathematical consensus and dismotivation is not simply a quantitative change. It is qualitative: consensus and dismotivation have evolved from a posteriori features of historical mathematical languages into an analytic characterization that renders a science mathematical. Just as mathematical statements were dismotivated from descriptions to standards of reference, the same happened to the very characterization of mathematics as formally consensual and dismotivated—it turned from a characterization into a defining criterion. There's no wonder that mathematicians reach higher levels of consensus over the validity of arguments compared to any other kind of scientists. Whenever a scientific language is formal enough and dismotivated enough to allow such levels of consensus, it becomes mathematical by definition.

But I have to qualify again. The fact that proofs *can* be broken down into pieces, each of which (but not all at once) *can* be more or less formalized by a trained professional within a reasonable amount of time does not mean that formalization is *actually* obtained. It also does not mean that all mathematicians know how to formalize all proofs, or that full research articles can be completely formalized with a reasonable amount of time and effort. Mathematicians usually argue and persuade each other by much less formal means. However, when a disagreement persists, it is usually possible to pinpoint it to a specific subargument, and resolve it by partial formalization.

This formalization is the last arbitrator. Like a Supreme Court, it is rarely appealed to. Still, every once in a while, in order to sort out a dispute, one must partially formalize a problematic chunk of an argument in order to convince critics that it works. Yielding supreme arbitration to formalization is what makes a scientific language mathematical, but it does not necessarily dominate mathematical practice.

And despite the power of formalization to promote consensus, it does leave some controversies undecided. Examples include Pierre Cartier's claim (Krömer et al. 2009) that the combination of set theory and category theory lacks a consistent foundation, Claude Rosental's (2008) documentation of a disputed theorem on fuzzy logic, as well as some other much more mundane and "low profile" disagreements between practitioners.

Interpretation

The stronger the authority of the formal-syntactic arbitration procedures, the more semantically fluid the language becomes. Indeed, as empirical and other procedures lose their power to arbitrate mathematical disputes, the semantic interpretation of mathematical claims is less committed to any specific setting, and becomes more and more underdetermined.

Now, one should not conclude that mathematics has no semantic purport. Nothing could be further from the truth. However, the semantic purport of mathematics is characterized by forms of indetermination and fluidity that are hardly ever associated with mathematics, because of the strong impression made by the rigidity of its formal syntax and consensus.

I'm not referring here to Hilbert's formalist claim that in axiomatized geometry we can arbitrarily replace "point, line, and plane" by "table, chair, and beer mug." This arbitrary replacement is precisely what doesn't happen in mathematical practice. Mathematical practice always has to do with some sort of semantic purport, but this semantic purport is not an objective reference—it is an open-ended process, as argued, for example, in William Thurston's (1994) discussion of interpretations of derivatives and of the importance of developing such interpretations in mathematical practice.

Before I explain what an open-ended process of interpretation or semiosis means, we should recall that "explanations come to an end somewhere" (Wittgenstein 2001, §1). Note that Wittgenstein makes this statement already in the first item of his *Philosophical Investigations*, long before one would expect a philosopher to make such a foreclosing statement (I owe this observation to Anat Matar). But he is right. It is a necessary fact that explanations come to an end. When I act on a statement, I impose a cut on the process of interpretation. And yet, it is also just as true of explanations that they can be resumed.

Here's how Peirce puts it:

> A **Sign** is anything which is related to a Second thing, its **Object,** in respect to a Quality, in such a way as to bring a Third thing, its **Interpretant,** into relation to the same Object, and that in such a way as to bring a Fourth into relation to that Object in the same form, **ad infinitum.** If the series is broken off, the Sign, in so far, falls short of the perfect significant character. It is not necessary that the Interpretant should actually exist. A being **in futuro** will suffice. (Peirce 1931–58, vol. 2, §92)

A sign needs an interpretant, and each interpretant becomes itself a sign of the same object. One often finds misinterpretations of Peirce's definition of signs that shift the object together with the interpretant (usually based on the confusing formulation in vol. 2, §303). However, these readings are true in a weaker sense: the object too belongs to its own semiotic chain. Indeed,

> The object of representation can be nothing but a representation of which the first representation is the interpretant. But an endless series of representations, each representing the one behind it, may be conceived to have an absolute object at its limit. (Peirce 1931–58, vol. 1, §339)

As the object tends toward its absolute limit and the interpretant gives rise to a potentially infinite series of interpretants,

> [a] symbol, once in being, spreads among the peoples. In use and in experience, its meaning grows. Such words as **force, law, wealth, marriage,** bear for us very different meanings from those they bore to our barbarous ancestors. (Peirce 1931–58, vol. 2, §302)

The metaphysical divide between optimists and pessimists of scientific rationality can often be drawn around the question of the purported objective limit. The transition from Peirce's symbol to Derrida's iterable sign is partly captured by giving up the objective limit. Indeed, Derrida's notion of sign is defined by its ability to be repeated:

[T]here is no word, nor in general a sign, which is not constituted by the possibility of repeating itself. A sign which does not repeat itself, which is not already divided by repetition in its "first time," is not a sign. (Derrida 1978, 246)

Now by dint of this fact of constitutive iterability of signs,

[e]very sign ... can be cited, put between quotation marks; thereby it can break with every given context, and engender infinitely new contexts in an absolutely nonsaturable fashion. This does not suppose that the mark is valid outside its context, but on the contrary that there are only contexts without any center of absolute anchoring. This citationality, duplication, or duplicity, this iterability of the mark is not an accident or anomaly, but is that (normal/abnormal) without which a mark could no longer even have a so-called "normal" functioning. What would a mark be that one could not cite? And whose origin could not be lost on the way? (Derrida 1977, 12)

I won't try to convince skeptical readers to go all the way here with Derrida. I will try to illustrate mathematical semiosis, and leave it to the reader to decide whether Derrida or Peirce capture it better. But before I do that, one claim needs to be qualified: the fact that a sign *can* break with every given context does not come for free. It doesn't just happen. It requires work. Someone has to cite the sign, reinterpret it, rewrite it, and disseminate it into new contexts. The outlets for such dissemination are limited and subject to competition, and sometimes gatekeepers or censors try to restrain the wild dissemination of signs beyond some preferred contexts (this is the notion of *rarity* in Foucault 2002, 133–41). So the work of decontextualization requires effort and force (the long and tortuous history of attempts at—and resistance to—reading some of Euclid's geometric theorems as arithmetical theorems is a case in point; see Corry 2013). It's not an arbitrary game without constraint.

To show how shifts of meaning may take place in mathematical practice, I'll make up an example involving iterated mathematical signs in a simple mathematical context. This is not quite a "real life" example following an actual historical or classroom development. This example is tailored for the purposes of this presentation. More realistic examples are brought in the following chapter and in chapter 6. Other examples that can be read in a similar light are available in Grosholz's (2007) deep and undercited account of "productive ambiguity" and in Fisch's insightful work on British algebra (forthcoming). Byers (2007) provides a treatment of ambiguity directed at a more general readership.

The mathematical signs we're going to consider here are 2-by-2 matrices. These are arrays of four numbers ordered in two columns and two rows, according to the pattern:

$$\begin{pmatrix} a & b \\ c & d \end{pmatrix}.$$

There are many things that a matrix can be interpreted as standing for. One such object is a parallelogram in a Cartesian plane. The preceding matrix can be used to describe the parallelogram whose vertices have the coordinates and $(0,0)$, (a,c), (b,d), and $(a + b, c + d)$. It can also be used to describe a linear transformation (a combination of rotating the plane around $(0,0)$, stretching it in various directions, and possibly reflecting it). Specifically, this matrix represents the linear transformation that sends the point $(1,0)$ to (a,c) and $(0,1)$ to (b,d). In order to simplify the presentation, I will assume that the matrices are positive orthonormal—that is, that the parallelograms they represent are unit squares, and the transformations are just rotations around the origin.

So far we have polysemy: several references for the same sign. The same matrix stands for a square and a rotation. But the fact that a signifier has several interpretations is less than interesting here, unless these interpretations interact. In order to get to the point of interaction, let's consider matrix products—namely, the operation:

$$\begin{pmatrix} a & b \\ c & d \end{pmatrix} \cdot \begin{pmatrix} w & x \\ y & z \end{pmatrix} = \begin{pmatrix} aw + by & ax + bz \\ cw + dy & cx + dz \end{pmatrix}.$$

Matrix product is an operation that takes two matrices and yields another matrix. What's important for us about matrix multiplication is that if X is a matrix that stands for a square, and Y is a matrix that stands for a rotation, then $Y \cdot X$ stands for the square you'd get by applying the rotation represented by Y to the square represented by X.

At this point, our two interpretations (square and rotation) relate to each other and interact. Mathematical practitioners interpret signs in different ways, and compose these interpretations. One can introduce many such compositions. For example, on top of the previous rotation-applied-to-square interpretation of products, one can interpret the product of matrices as a composition of the rotations that they represent (the product of a matrix that stands for a 45-degree rotation and a matrix that stands for a 30-degree rotation will then stand for the combined 75-degree rotation). Both these interpretations are useful in practice. But for the purposes of this motivational story, let's assume that we have in mind only the rotation-applied-to-square interpretation.

So far, our two interpretations for matrices (rotations and squares) coexist; they do not compete with each other or work against each other. So let's look at one more thing that we can do with matrices: raising them to the second power. The formula is:

$$\begin{pmatrix} a & b \\ c & d \end{pmatrix}^2 = \begin{pmatrix} a^2 + bc & ab + bd \\ ca + dc & cb + d^2 \end{pmatrix}.$$

We can think of this operation as a function that takes a matrix standing for a square, and yields another matrix standing for another square. We can check and verify that the angle between the resulting square and the x-axis is twice the angle between the original square and the x-axis. So far, we're just adding more operations and interpreting them in coherent ways.

But what happens when we put things together? For instance, what happens if we note the simple fact that $X^2 = X \cdot X$? Here things start to jar. On the left-hand side, our interpretation has to do only with matrices interpreted as squares. We don't need any other interpretation to read the left-hand side. But when we apply our interpretation to the right-hand side, we are forced to take the sign X, which on the left-hand side stood alone with a single intended interpretation (square), and impose upon it another interpretation, that of a rotation; otherwise, the right-hand side wouldn't make sense in terms of the interpretations already available to us.

The single left-hand side X splits into two at one and the same time. Here we don't just have polysemy. By setting this equality and retaining our previous interpretations, we force a shift of meaning: a matrix designating a square suddenly designates a rotation, because of a formal equality that happens to hold. A formal manipulation combined with inherited interpretations forced a shift of meaning: from one sign and one interpretation, we turn to a reiterated sign and two interpretations. We can no longer think of the left-hand X as standing only for a square, as we could before, because the right-hand side and the connecting equality sign force our second interpretation on X. The left-hand interpretation was "contaminated" by the excess meaning in the right-hand interpretation. One interpretation was forced on another by following a formal identity.

But we have to qualify in what way this shift of meaning is forced. Of course, we needn't have acknowledged any of the interpretations I have suggested earlier. One can do matrix algebra with many other interpretations, including an interpretation that views matrices simply as arrays of four numbers. However, here we're dealing with mathematical practice, and in practice we always work with some of the inherited interpretations that signs accumulated on their way to us.

Mathematics is useful and interesting because it is interpreted. In saying that, I am referring not only to interpretation for the purpose of application, but also to interpretation for the purpose of generating mathematical conjectures and proofs. I know no mathematician who never interprets her or his symbols when thinking about mathematical problems. I know no mathematician who sticks to just one interpretation.

In fact, mathematicians often work at one and the same time with formal interpretations (the matrix as a purely syntactic object) and several semantic interpretations (such as squares and rotations). This ambiguity is one aspect of the term "analytic a posteriori": mathematical claims are ambiguously analytic *and* committed to a posteriori experience. The two interpretations do not come one after the other; they often intertwine.

Whenever one has an isomorphism, one has (at least) a double interpretation: one in terms of the domain of the isomorphism, and one in terms of its range. My point is that mathematical interpretations can and do bump against each other to force shifts of meaning. Going from the left-hand side to the right-hand side of an equality, a sign may change its interpretation. And this holds not only for specific signs, but for entire domains of knowledge as well. Analytic geometry, for example, is not simply two independent ways of thinking put together—geometric and algebraic. It is a novel geometric-algebraic way of thinking, which is historically distinct, and not reducible to a disjoint union between classical geometry and symbolic algebra (more on that in the first section of chapter 6).

But I want to make things more involved. I want to show how interpretations strike limits. For that purpose, consider one more matrix operation: transposition, denoted by a superscript T. Its definition is simply:

$$\begin{pmatrix} a & b \\ c & d \end{pmatrix}^{\mathrm{T}} = \begin{pmatrix} a & c \\ b & d \end{pmatrix}.$$

This operation is easy to interpret in terms of squares and rotations. If a matrix stands for a square, its transpose stands for the square obtained by reflecting the sides of the original square around the main axes. If a matrix stands for a rotation, its transpose stands for the same rotation in the opposite direction.

Now there's an easy theorem stating that $(X \cdot Y)^{\mathrm{T}} = Y^{\mathrm{T}} \cdot X^{\mathrm{T}}$. Let's try to read this theorem in the terms of our preceding interpretations. If we think of Y as a square and of X as a rotation, then the left-hand side makes sense (apply rotation X to square Y and then reflect its sides), but the right-hand side involves multiplying a square to the left of a rotation, which is not something we've considered so far. (In general, matrix multiplication is not commutative, so we can't just switch the matrices around.) If, on the other hand, we read X as a square and Y as a rotation, then the left-hand side no longer makes sense.

One way to deal with the issue is to offer another interpretation for transposition. For instance, we can stipulate that if X is read as a square, we should read X^{T} as a rotation, and vice versa. Then, if we decide that X is a rotation and Y is a square, the left-hand side product $X \cdot Y$ yields a rotated square, while on the right-hand side of the equality, Y^{T}, a rotation, stands to the left of X^{T}, a square, and we can maintain the interpretation of the product to yield, again, a rotated square.

Both sides now make sense. But given this interpretation, the equality between the two sides no longer makes sense. Indeed, on the left-hand side, we have a transposed rotated square, which, according to our interpretation of transposition, must stand for a rotation. But the right-hand side is simply the application of a rotation to a square, and is, according to our interpretation, a square. We obtain a situation where a rotation on the left equals a square on the right, which is likely to appear more objectionable than the simple claim that a single matrix can represent either. In a way, what we saw is that mathematical syntactic truths move faster than semantic interpretations, and sometimes leave them behind.

When confronted with this kind of interpretive dead-end, we can react in various ways. One reaction is to seek other interpretations for transposition and multiplication that work coherently together, and at the same time allow us to retain a sense of rotating squares for application purposes. This would be a "reconstruction of interpretations." Another attitude is to keep using our interpretations locally—that is, to change our interpretations of multiplication and transposition as we go along, even if it means that X or transposition is interpreted in more than one way across a single sign of equality. One can refer to this as "superposing interpretations." Another strategy is to set interpretations (in terms of squares or rotations) aside for a while, and bring them up only opportunistically, in specific locations where they are actually useful. This might be called "deferring interpretation."

All these approaches have productive roles in mathematical practice. Mathematics is practiced through and across reconstructions, superpositions, and deferrals of interpretations. These processes never come to an end, because formal manipulations never come to an end. Things only get more and more involved, and the previous example traces only a few initial steps in a web of semiosis.

There's yet another strategy, of course, more respectably philosophical: to seek an all-encompassing ontological grounding or logical reconstruction. I doubt that this can be a successful effort, but the important point is that mathematics does not *require* a global grounding. Mathematics works in practice across, against, and in conflict with locally reconstructed, superposed, and deferred interpretations.

There are many mathematical languages where handling conflicting interpretations is a challenge that has to be dealt with carefully in order to avoid spinning into senselessness. However, as the arbitration over the validity of mathematical arguments is increasingly assigned to a syntactic formalization, the need to manage interpretations coherently becomes less urgent, and semiotic fluidity is subject to ever diminishing constraints. This is how syntactic rigidity supports semantic fluidity.

But what happens when we leave mathematics? Can't we say at least that at the moment of extra-mathematical application, there is a definitive collapse into a single interpretation? Well, suppose a programmer used a certain matrix to designate a certain square, to be displayed and rotated on the screen. Has interpretation come to an end? Not quite. The code is to be processed by a certain compiler, then executed on a certain machine, and eventually output onto a certain medium. Anyone experienced with the endless machine-specific variations that can ensue is well aware that interpretations have not yet come to an end.

And when we finally observe the square rotate on some LCD screen, has interpretation now finally come to an end? It has not, at least not as long as someone is there to observe the rotating square and interpret its movement: experience it aesthetically, derive information from the display, act on whatever the rotating square prompts them to do, and so on. And things needn't end there. This experience may be remembered, recalled, evaluated, recounted, recontextualized; it may instruct us, reproduce itself in future experiences, enter chains of interpretive expectations and habits—in short, interpretations, like explanations, don't have a well-determined end. They must factually come to ends, but any given end is open to future resumption.

The preceding example assumes that we actually leave mathematics. In fact, as captured by the term "analytic a posteriori," it is never finally decided whether an interpretation carries a sign outside mathematics. When I interpret a matrix as a square, is the square no longer mathematical? It depends. A square on an LCD screen may well be an empirical object of observation. But it may just as well be a mathematical object.

Indeed, according to Wittgenstein "the sentence 'The figure I have drawn here … ' may be used either mathematically or non-mathematically" (1976, 117). For Wittgenstein, whether the square is inside or outside mathematics depends on how we operate with it. If our dealings with the square set standards (say, if we state that the square's diagonal divides it into two congruent parts, regardless of what empirical measurements of the LCD screen suggest), then the square is still mathematical. If our dealings with the square are more empirical or pragmatic (say, we actually measure the two halves of the square and proceed based on the result of this measurement, even if it suggests a discrepancy), then, according to Wittgenstein, we're no longer inside mathematics.

There's nothing in the square itself that forces us to use it either way, that forces us in or out of mathematics. Our interpretation may oscillate in and out of mathematics in ways that might question the topology of these in/out relations. Our understanding of the square can remain undecidedly analytic and a posteriori. Mathematical interpretations do not necessarily mark clear boundaries of final interpretations or ways out of mathematics.

This open-endedness is not the margin of mathematics. It is its de-centered center, its characterization as a combination of semiotic dismotivation and syntactic validity arbitration. Even if mathematics is considered as geared not toward practice, but toward some ideality, a mathematical practice that does not involve reinterpretation is a practice that is bound to become outdated, once a given interpretation is rendered obsolete by traveling—historically, culturally, intersubjectively—across our evolving life worlds.

Reality

The debate concerning the reality of mathematical objects is organized around two main arguments. The basic objection to the claim that mathematical objects exist as independent abstract entities is the problem of our spatio-temporal access to abstracta. The main objection to the rejection of the reality of mathematical objects is the problem of how these made up entities become involved in applicable knowledge.

Lately, two interesting arguments have been put forth that reconfigure the problem. Emily Grosholz (2014, forthcoming) suggested that even if we had full access to mathematical objects by means of a "maths genie," we'd still be unable to account for our mathematical knowledge. She imagines Fermat asking the genie for the proof of his famous last theorem. But what is Fermat asking for precisely? He obviously can't ask for Wiles's proof of the Taniyama-Shimura conjecture, because he can't even imagine the mathematical objects involved. If, on the other hand, he's asking for something more open ended, how would the genie respond? Which of the many possible mathematical paths to a proof of Fermat's last theorem would he present? And which criteria of validity should he adhere to, given the gap between Fermat's criteria and contemporary ones? Having access to the totality of mathematical objects, relations and criteria of validity, would he have to produce this unbounded mass as a whole?

Grosholz's point is that our mathematical knowledge depends on our historical and contingent formation of mathematical problems, not just on access to mathematical objects. Even if we believe in the independent reality of mathematical entities, our mathematical knowledge is a tentative and underdetermined exploration of an evolving mathematical landscape. The fact that two expeditions climb the same really existing mountain and reach the same really existing peak doesn't mean that they travel the same trail, see the same view, share the same experience, or interpret it in the same way. To explain mathematical knowledge, we need to account for the historicity of problems, representations, and interpretations, and not just rely on the fact that mathematical objects are out there and can be somehow accessed.

The second argument comes from Penelope Maddy (2011, ch. 4, §2). She points out that the problem of mathematical applicability is not solved by the reality of mathematical objects. Early modern mathematics was constrained to a large extent by its relation to physics. Today, this relation has lost its formative status, as mathematical development is constrained by intra-mathematical requirements for mathematical richness and productivity (Maddy 2007). The unexpected emergence of empirically applicable results from a quest motivated by intra-mathematical concerns is not reducible to the objective reality of the mathematical objects discovered along the way. Indeed, the fact that something is real doesn't explain its emergence from the internalist constraints of a discourse that's foreign to its context of empirical application. (For Maddy's own explanation of applicability, see her 2007, 329–43; my take is available in the closing section of this book.)

So in Grosholz's terms, even if we solve the realist access problem, we have not given a proper account of how human mathematical knowledge works. In Maddy's terms, even if we assume the reality of mathematical objects, we have not given a proper account of mathematical applicability. The actual tasks of realist philosophers of mathematics turn out, then, to be surprisingly close to those of anti-realists: explain how mathematical knowledge evolves, and how the intra-mathematical constraints on its evolution produce results that are relevant in seemingly unrelated empirical applications.

It is interesting to note that Grosholz and Maddy are both friendly to some versions of realism—they both reject the view that mathematicians simply make things up as they go along, independently of objective mathematical constraints. It is also important to note that their articulations of mathematical existence and truth focus on empirical science as their point of reference.

As a result of this reference point, when Maddy considers the existence of mathematical objects and the truth of mathematical statements, her models are empirical scientific existence and truth. She compares and contrasts the older, empirically constrained mathematical methods and norms to the intra-mathematical quest for depth and richness that in her view constrains contemporary mathematics. She finds that a philosopher who embraces a holist and continuous approach to science and mathematics is likely to extend the empirical notions of existence and truth to contemporary mathematical objects and theorems. A philosopher who emphasizes the difference between the older and newer mathematical methods and norms, on the other hand, may rule that empirical scientific notions of existence and proof are distinct from those of mathematics.

For Maddy, the choice between the two options is a more or less conventional decision about the extension of old terms to a somewhat related new context—a context not covered by the old terms' original articulation, but not entirely disjoint. The concern over existence and truth in contemporary mathematics is therefore not about verifying whether the old meanings of these philosophical terms apply to contemporary mathematics; rather, it is about deciding whether we should enrich these old meanings so as to conform with contemporary mathematics, or confine them to their original context.

But existence and truth do not derive their meaning only from the analytic philosophy of empirical science. If we want to confine ourselves to the context of empirical scientific existence and truth, we might as well ask whether mathematical objects and claims are "scientific," rather than whether mathematical objects "exist" and mathematical claims are "true." If, on the other hand, we do not wish to focus on the context of natural science, we should shift our attention from empirical existence and truth to other manifestations of reality. This is where I'm heading next.

Constraints

It is obvious that mathematics as a social endeavor is real. Beyond the debated existence of mathematical objects and truth of mathematical claims, there lies the undisputed reality of mathematical institutions and practices: university departments, mathematicians, journals, publications, interdisciplinary collaborations, mathematical modeling, conferences, professional societies, grants, teaching, media coverage, popular representation, and the politics of research and higher education. All these things make mathematics very real. (One might even go as far as saying that if one is willing to accept objective mathematical abstracta, then the social abstracta that might be involved in the preceding list could serve as an ontological means to access mathematical abstracta!)

The reality of these mathematical institutions and practices imposes strong constraints on mathematical objects and statements. Indeed, in the first couple of sections of this chapter we saw how classical mathematical objects and statements function in the span that stretches from empirical descriptions to standards with respect to which we describe things—the span subsumed under the titles "dismotivation" and "analytic a posteriori." Functioning within this epistemological span imposes constraints on mathematical objects and statements, as they have to fit into it.

Contemporary mathematical objects and truths are further constrained by the arbitration of partial formalization, which, as we saw earlier, plays a key role in enabling mathematical consensus and the fluidity of its interpretations. Indeed, mathematical knowledge should be translatable into a family of formal languages that get real existing mathematicians to agree on whether one thing follows from another. These languages, in turn, are constrained by the sociological and cognitive human capacities to be trained to follow only certain kind of rules in very similar ways. This constraint significantly narrows the kind of claims that mathematics can state.

Mathematical objects and truths are also subject to the constraints of interpretability and representability (explored earlier in this chapter) that allow real existing mathematicians to actually explore them, as this exploration is never reducible to purely formal languages. Mathematical practice depends on reconstruction, superposition and deferral of interpretations. The conflicting interpretations involved are not (only) a logical impediment, but (also), as we noted earlier, a productive force in mathematics. These translatability and representability impose further constraints on mathematical objects and statements.

Moreover, contemporary mathematical objects and statements are constrained by informal notions of maximizing mathematical productivity and depth, similar to those invoked by Maddy. This is a very real constraint. Maddy believes, for example, that this kind of constraint is likely to guide mathematicians to discover and endorse new mathematically productive axioms that will eventually decide such statements as the continuum hypothesis, which is independent of the current standard axioms. This last prediction seems to follow from Maddy's focus on a certain school of set theorists, but generally related intramathematical productivity constraints hold in a much larger context.

Another constraint has to do with the web of inter-communal cross-domain mathematical relations. Even in mathematical cultures where consensus subsists, a group of mathematicians that would present objects and claims in ways that do not interact in interesting ways with those of other mainstream groups is not likely to be published or cited. Their mathematics is not likely to evolve and leave a mark. As Stav Kaufman (2016) put it in a highly sober case study analysis, "the community's response takes part not only in the selection of problems, but also in the shaping of the results themselves," in the sense of articulating them in terms relevant to different mathematical communities. (For a more general conception of mathematics as a social phenomenon, see Ernest 1998.) This constraint therefore imposes its own limit on mathematical evolution.

It is important to note that the considerations that constrain mathematical development cannot be neatly separated into institutional versus purely mathematical. The unbounded span of formally correct assertions is meaningless as an undifferentiated whole. Mathematical objects and statements make sense only when interpreted and organized with respect to other mathematical, worldly, and human entities. This organization of knowledge can be read in terms of the intramathematical impact of some statements and objects on our access to the formation and proof of others (an old-school history of mathematics), in terms of the open-ended historical dialectic of mathematical problems (Cavaillès 1994; Lakatos 1976), and in terms of interpretive negotiations over mathematical meaning between different groups of mathematicians (social construction of knowledge). Either way, the evolving organization of knowledge imposes strong constraints on what can surface as a mathematical issue.

But the social infrastructure goes even deeper. Historians and sociologists of mathematics teach us that contemporary mathematical objects and claims are constrained by systems of exchange, institutional hierarchies, and material means of production that enable mathematicians to critically assess each other's efforts and reach agreement concerning validity. These economic constraints and their impact are analyzed from a game theoretic point by Paul David, who suggests that changes in early modern structures of funding for scientists transformed scientific culture from secretive and conflicted to more open and consensual (David 2004, 2008; Høyrup 2008 supplies an interesting case study in the context of the abbacus culture). The contemporary working of mathematical communications is documented in anthropological investigations such as Rosental (2008) and Barany and MacKenzie (2014). Without such conduits and constraints, mathematics may return to be just as polemical as it had been in many of its historical manifestations—polemical not only about what's important or interesting, but also about what is mathematically valid, producing a lot of what (from our point of view) would be odd mathematics. Such polemic had significant impact on the content, role, and position of mathematics in various historical systems of knowledge.

Social and political contexts affect mathematical knowledge in other ways as well. In the first chapter, we examined the controversy over infinitesimals with its religious and authoritarian overtones (Alexander 2014). Today, grant-giving institutions (national funds, military research funds, private funds, corporations) can influence the internal organization of mathematics. This does not mean that some conspiratory cabal makes up mathematical knowledge according to its political whim. It does mean that the many constraints imposed on the evolution of mathematical knowledge cannot be disentangled into purely internal versus social.

One might try to distinguish constraints that are forced on us no matter what ("If I jump off this ledge, I'll crash onto the ground") from ones that depend on society or ourselves ("If you don't speak our language, no one will understand you," or "I must do what I believe is morally right"). But, in fact, all these constraints can be circumvented: ("If I jump off this ledge, I'll crash onto the ground, *unless I grasp onto something*"; "If you don't speak our language, no one will understand you, *unless you get an interpreter*"; "I must do what I believe is morally right, *unless I'm willing to deal with the guilt and consequences*"). The only forces that are "truly" inescapable are those that have been dismotivated and rendered analytic a posteriori: I can't escape gravity or the fact that $2 + 2 = 4$, because whenever I come up with a clever setting where they seem to have been escaped, adherents of arithmetic and physics can dismiss my counterexamples by saying that I did not understand correctly their empirical purport. (I discussed arithmetic examples at the beginning of this chapter; for the context of the laws of physics, see Cartwright 1983.) The continuum of natural-social constraints is real, regardless of its internal divisions.

In fact, all the historical-philosophical problems reviewed in chapter 1 can be seen as constraints: tying mathematics to natural phenomena or to independent thought, designing math as a *sui generis* source of knowledge or reducing it to others, allowing it to entertain monsters or forcing it to do away with them, aligning it with this or that authority—these are all competing constraints that mathematics engages as it evolves in different settings. Adherence to one constraint over another reflects a certain cultural preference that shapes mathematics in its image.

The point is that the continuum of constraints that form mathematics is real in that it sets limits to what can emerge as mathematical. In fact, the conjunction of these constraints is so strong that sometimes only a single possible result can successfully negotiate them. Mathematics is the result of organizing and prioritizing these and other constraints in a manner that establishes a viable system of knowledge that can survive in given times and places. And different times and places can yield different real mathematical cultures, concepts and claims.

When viewed from this perspective, mathematical reality is comparable (and contrastable) with the reality of many other branches of knowledge, including those that are not answerable (or fail to answer) to scientific norms and standards. Maddy's favorite examples are astrology and theology, and I would add my own heretic creed: poststructural philosophy. Such institutions of knowledge are very real, and their postulated objects and statements are highly constrained, even when these constraints are not continuous with those that shape empirical scientific existence and truth, and even where they leave practitioners with significantly more freedom.

The following should be considered trivial, but this triviality is not reflected by current philosophy of mathematical and scientific knowledge: no institution of knowledge simply makes up things as they go along. The popular image of social construction as a bunch of people coming up with some mutually agreed mumbo-jumbo is nothing but a pastiche. (The satirical website *The Onion* had a great piece mocking this absurdity, titled "Historians Admit to Inventing Ancient Greeks.")

Like mathematics, institutions of knowledge make projections and provide guidance to people in ways that are integrated into their forms of life. As in mathematics, these projections and guidance are not always empirically successful, and the various branches of knowledge have to account for these failures. These accounts can't be too general or too trivial. Indeed, if theology or mathematics could only justify themselves by such generic claims as "divine justice is served—if not here, then in the afterlife" and "2 + 2 = 4—if not here, then in the abstract realm of number," then we might take theology or mathematics less seriously. We could claim that bad deeds are punished simply due to social convention, or that 5 dollars and 7 dollars are 12 dollars simply because that's the result imposed by the algorithms implemented and forced on us by banks. But if theology and mathematics survive, it is because they can sometimes provide more compelling accounts for their successes and failures and cast more robust nets of sense making coordinates.

A doctrine may be invented or initiated by a single person, but the real existence of a distributed branch of knowledge cannot be reduced to such an authority, even if it wields strong repressive powers or manifests exceptional charisma. Branches of knowledge that survive are highly constrained by a complex network of material circumstances and cultural roles. They are entangled into politics, social psychology, and empirical reality. An incredibly complicated conjunction of realities must be present in order to create, disseminate, and maintain a branch of knowledge that really exists over large spans of space and time—be it mathematical or theological knowledge.

One of mathematics' most impressive and seemingly unique features is its consensus, which makes it seem more real than other branches of knowledge. But in order to evaluate this consensus, imagine a largely consensual group of theologians that have their best students sent to establish mutually independent subsidiaries. Are the different schools likely to maintain consensus, or are they likely to establish divergent tenets? How would these theologians fare compared to mathematicians in a similar situation?

The knee-jerk reaction of a person who, like me, grew up in a secular pro-scientific environment is to claim with certainty that the mathematicians are much more likely not to conflict, even if their research goes in very different directions. But my point is that this consensus cannot be derived from the truth of mathematical claims, even if such truths were out there. Just because something is out there, doesn't mean we're going to get it. Mathematical consensus has to do with the constraints surveyed in this chapter. If the mathematical subsidiaries in our hypothetical situation would weaken the system of dismotivation, formal arbitration, and fluid interpretation, if their material conditions would preclude certain structures of communication and exchange, if their subsistence would depend on a different economic arrangement, they may become very polemic. In such cases, we may say that they're no longer doing (good) mathematics, but that just reflects our analytic a posteriori definition of what mathematics means (namely, that mathematics is a discourse that achieves consensus by relying on the supreme arbitration of formalization according to rules that can have people trained to follow them consensually). That does not reflect, however, the many historical cultures that we recognize as engaged in mathematical practice due to their historical continuity and family resemblance with respect to our own mathematical culture.

The bottom line is that general questions concerning the existence and truth of objects and statements carry notions of existence and truth, which reflect the constraints of some given domain, into different contexts subject to different constraints, in ways that make our answers rather contingent. Instead of trying to pull the conceptual blanket of existence and truth across too many domains, the philosophy of mathematics might want to study the specifics of the contingent pools of constraints that make mathematical objects and statement non arbitrary and sometimes even uniquely determined.

This approach would allow us to compare mathematical truth and objectivity constraints with those imposed on other, not necessarily scientific, branches of knowledge. As a result, this kind of analysis would be much more useful for our evaluation of the unique position and merits of mathematics inside our institutional system of knowledge.

Relevance

Let's consider the problem of mathematical reality from another angle. Our starting point will be Putnam's (2004) version of realism. In this version, Putnam rejects the referential objectivity of mathematical entities and allows for the relativity of truth with respect to linguistic articulations. For example, what counts as a single object in a certain language might not count as an object in another, and so the existence of objects may turn out to be relative to the specific language used and to schemes of translation.

Still, according to Putnam, truth goes beyond the rules of articulation and derivation in specific languages. He acknowledges a register of "conceptual truths"—namely, those for which "it is impossible to make (relevant) sense of the *assertion* of [their] negation" (Putnam 2004, 61). Putnam's examples of conceptual truths include simple logical truths such as statement that fit the template $p \rightarrow q \equiv (\sim q) \rightarrow (\sim p)$ and mathematical statements such as $2 + 2 = 4$.

To clarify that his notion of conceptual truth embraces fallibility, Putnam insists that these truths can be refuted by coming up with a (relevant) context in which their negation does make sense. His example (2004, 61–62) is how Riemannian geometry debunked some of the elementary truths of Euclidean geometry (for example, the sum of angles in a triangle equals two right angles) from their status as conceptual truths by giving sense to their negations.

The catch is, I believe, the bracketed term: "(relevant)." What about all those cases, discussed in the opening section of this chapter, where elementary arithmetic doesn't hold empirically (water drops, very large numbers)? What about addition modulo 3, where $2 + 2 = 1$? Why are these exceptions not relevant, whereas Riemannian geometry is? More precisely, for whom, and in which contexts, is the latter relevant while the former is not?

One could argue that light rays were thought to obey the laws of Euclidean geometry, and then turned out to obey the laws of Riemannian geometry. This produced an interpretation of the sum of angles in triangles that negated the Euclidean interpretation, and was clearly relevant for some scientists, and indirectly for many people who use advanced technologies such as GPS. On the other hand, we can't claim that we used to think that two apples and two apples made four apples, but later found out that they make five in some scientifically relevant context. There have always been things that didn't add up in line with $2 + 2 = 4$, but this is not what Putnam means by "relevant," because his articulation of relevance is restricted precisely to the kind of context where $2 + 2 = 4$ has been considered true.

However, as we saw at the beginning of this chapter, the context for which basic arithmetic is considered true is not well defined in advance, and so we risk a circular definition of "relevant" as referring to those, and only those contexts, where basic arithmetic keeps holding (instead of the contexts that obey a certain set of requirements that seem to guarantee arithmetical truths). This approach would confine us to an infallible analytic a posteriori realm, which is not what Putnam is after.

In reality, different contexts would be "relevant" for different people in different times and places. Indeed, for many people light rays and GPS are not relevant at all; for others weird ways of counting may become relevant, as they suddenly figure out, contrary to what they have always believed, that in the context of managing their business, 2 + 2 may turn out equal to 3.8 due to the law of diminishing returns or bulk purchase discounts.

I believe we have no articulation that specifies in advance what would be relevant and what would not. Yet it is crucial that Putnam's conceptual truth hinges on a notion of relevance. It is crucial, because it means that different articulations of relevance may commit us to different articulations of truth.

Suppose, for example, that we accept Wittgenstein's claim that mathematical statements are constrained by their role as standards with respect to which we describe, rather than as descriptive statements. In that case, "relevance" is not about the existence of an empirically descriptive interpretation, but rather about integration into a system of standards with respect to which we describe empirical observations (as, for example, the Black-Scholes formula from our opening vignette is a relevant standard for practitioners without necessarily being a very successful empirical description of option pricing). As our host of empirical descriptions change along with our changing instruments and representations, so can old mathematical standards of description become obsolete, and others become relevant.

Maddy's account takes us in a different direction. According to her view, mathematical objects and statements are constrained by the intra-mathematical goal of enriching and deepening mathematics. So again, "relevance" is not about an empirical interpretation, but rather relates to the goal of mathematical productivity.

So following Putnam, but considering the overall web of constraints imposed on mathematics, the truth of a mathematical statement turns out to hinge on the relevance of its interpretations (or the interpretations of its negation) with respect to the different constraints that shape a given mathematical culture—for example, empirical adequacy, standards of description, internal productiveness, or any of the constraints pointed out in the previous section. But the bottom line is that the truth of a mathematical statements hinges on its relevance to a nonmonolithic set of constraints.

To appreciate the force of this articulation of truth, take the following counterfactual example. Imagine an intelligent marine culture that lives in areas of very strong currents. For this culture, hydrodynamics is the most relevant mathematical-physical reality. And, due to natural selection, calculating elementary hydrodynamic projections is as easy for these marine creatures as is counting on fingers for us (say, they literally generate small controllable currents with their limbs and feel the results of their combinations with their sense organs). Based on this intuitive capacity, they develop a sophisticated and practical hydrodynamic calculus.

On the other hand, living in an environment of strong currents and having their sense organs and cognition shaped by this environment, doing one-one correlations is practically impossible for them. For them, trying to correlate two sets of four objects (even in their minds) is as impossible as trying to correlate two large sets of flickering fireflies would be for us. They do not perceive reality as analyzable into discrete constitutive elements, but as gestalts of currents. As a result, these marine creatures have no discrete arithmetic in our sense of the word. They have very different notions of quantity, mathematics, and computability. (If you want to think of a less counterfactual example, think of animal sensing and cognition, but here I'd like to imagine some highly intelligent culture.)

Suppose the computational richness of their ways of doing math is comparable in scope (if not identical in content) with what we can model with our discrete mathematical logic. Perhaps they can even efficiently solve problems that we find intractable, and vice versa. (This wouldn't be unthinkable, given that the quantum computation model and some DNA computation models can solve problems that Turing computing would take too long to deal with, or, being less exotic, that for certain problems in certain size ranges, computing based on dedicated hardware is sometimes more efficient than implementation on an all-purpose digital Turing computer.)

Now suppose humans meet these marine creatures, and—being certain, as humans often are, that their way of seeing things is better—try to teach them elementary arithmetic. Since the marine creatures' senses are not designed to follow, it's probably going to be very difficult. Perhaps the humans may be able to do this indirectly, by providing the marine creatures with instruments that rely on arithmetic to make relevant predictions—for example, an instrument that counts and compares the number of food portions to the number of eaters in order to determine whether there's enough food—instead of the marine creature's indigenous practice of swimming around the food and the eaters and deriving their estimate from the resulting currents.

Maybe, based on the human's predictions and instruments, the marine creatures might come up with some tentative theories of what discrete numbers might be and how they might work, based on all sorts of turbulence models. (After all, if we can generate continuous models of discrete phenomena, perhaps they could too.) But perhaps not. It is most likely that humans won't be able to communicate on a high level with creatures who can't count discrete objects. Too much of our language and life form depends on that capacity.

In the latter case, it may be as true for the marine creatures that empirical one-one correlations have no stable results, as it is true for physicists that quantum states are not univalent before they're measured. The rejection of the notion of one-one correlations between sets of objects would probably survive indefinitely in their culture. The reality of discrete numbers and the truth of elementary arithmetic may simply not be *relevant* for these marine creatures. They may not be able to give our arithmetic a relevant and sensible interpretation (as, perhaps, we of theirs).

If this is so, then numbers will simply not exist for these creatures— at least as long as we impose on the term "exist" the criterion of relevant sense making. It is in this sense that the conceptual truth of mathematics depends on relevance. The same can work the other way round: their concepts and methods may be inaccessible to our constrained embodied cognitive capacities and technologies.

Of course, one could object that even if the marine creatures don't know it, arithmetic truths are relevant for them. Suppose that if any of these marine creatures ate 10 blowfish, they would die. Then it's obviously relevant for them that they shouldn't eat 4 blowfish, and then another 6 blowfish. So numbers and arithmetical truths such as "4 + 6 = 10" are relevant for them, even if they can't figure it out. But in fact, from the marine creatures' point of view, what's relevant is that the flows correlated with what they would articulate as *such bunch of blowfish* (which approximate what we'd call 4 blowfish) and then *such other bunch of blowfish* (approximately corresponding to our 6 blowfish) mark danger. Identifying these flows does not necessarily require a notion of numbers—it may depend on something unique to how water flows around blowfish. Our abstraction of a notion of number from collections of different kinds of object is not vindicated by the fact that one shouldn't eat 4 blowfish and then another 6. Just because we think in these terms, doesn't make them relevant for the marine creatures.

Now, let's turn back to humans. When assessing the pragmatic reality of mathematics, that is, how it makes relevant sense to humans, we must consider two facts. First, that inter-translatable forms of higher mathematics have evolved and disseminated among many distinct cultures whose elites were affluent enough to dedicate time to abstract knowledge. This means that mathematics makes sense on a more than local level. Second, we should consider that higher mathematics is extremely difficult to teach and is nearly ungraspable to many humans. Indeed, among the topics taught in schools, mathematics is probably the one getting the most teaching hours and least level of understanding, at least according to people's self assessment.

Mathematics is crucial for our technological life form, but is accessible in all but an elementary form to a small minority alone. Higher mathematics is real to most humans only due to their vaguely successful communication on the subject with the representatives of the culture of higher mathematics. I think these observations are important for estimating the kind of relevance that the reality of mathematics has in a human context.

We could claim that this last articulation is irrelevant, because for mathematics to be universally real, it is enough that some (actual? possible? human?) culture can attain it, regardless of unsuccessful students and counterfactual marine creatures. But then we risk bestowing too much reality on highly suspect forms of knowledge just because *some* bunch of weird people managed to hold on to them sometime, somewhere. Again, as earlier, if we restrict the term "knowledge" to that which is obtained by scientific methods and norms, this risk may be circumvented; but, again, if this is our path, we might simply ask whether mathematics obeys scientific norms, rather than confront it with ambiguous and dangerous terms such as reality, existence, and truth.

If, however, we prefer to follow the relativity of relevance suggested here, then existence and truth end up reflecting how specific constraints on knowledge formation are relevant to specific forms of life.

These constraints belong to a continuum that spans human and non human physical reality, cognitive capacity, practices with tools, and social constraints. The more relevant a certain system of constraints and the more force it applies on us—the more reality or objectivity or truth we can assign to the resulting objects and statements. Reality, objectivity, and truth become, then, relational concepts ranging over a continuum, rather than binary universals. Mathematics can be more real to one kind of life form than to another, and entirely unreal to a third.

These relational notions of reality are not simply made up on whim; they depend on constraints that wield a real force. Making up these constraints or doing away with them may be extremely difficult or entirely impossible for a given life form (or to any life form of which we can make relevant sense). Heading in this direction, the philosophical task would involve an analysis of how sets of constraints include or exclude interpretations that make relevant sense of some assertions to some forms of life. Philosophy may then be about renegotiating constraints in order to find new ways of making relevant sense, or analyzing the force of constraints that we can't successfully renegotiate or remove.

If we think in this way, then our judgment concerning instituted branches of knowledge shifts from epistemological and ontological grounding to pragmatic, ethical and political considerations. The main concerns then become: To what extent can we impose or escape a given system of constraints? Do these constraints yield something useful? Do they serve the common good? No matter how real mathematics (or, say, religion) is, the mathematization (or deification) of our culture and science imposes constraints on what we say and do, and we should judge mathematization (or deification) based on our capacity or incapacity to handle these constraints, and on the impact of these constraints on people's lives.

I can't rule out philosophical attempts to discuss constraints that would apply to a large set of life forms, or even to all (relevantly) imaginable life forms, yielding, in turn, some sort of more or less universal truths. But I'd rather we focus on more pragmatic considerations of those constraints that have to do with more concretely relevant settings, such as human mathematics.

Conclusion

The first chapter of this book narrated histories of the philosophy of mathematics around four formative axes. In order to wrap up the current chapter, I'll try to see how the conceptions presented here work in terms of the preceding narratives.

The first narrative contrasted natural order and conceptual freedom. Is mathematics committed to one or the other? The account of dismotivation and the analytic a posteriori is meant to reflect on this issue. Mathematical knowledge formation is a process that is sometimes committed to our empirical observations of natural order, but at other times shapes our understanding of natural order by setting standards for describing natural phenomena.

Still, as expounded in the sections on interpretation, consensus, reality, and constraints, empirical observation has no exclusive authority over mathematics. Mathematics develops its interpretations, arbitration mechanisms, and other constraints in relation to various social, practical, and cognitive circumstances, evolving away from natural application.

This does not mean that mathematics has become conceptually "free." Even if today most of the constraints imposed on new mathematics have little to do with empirical observation, these constraints are tangibly real. There's obviously still a large measure of contingency in the evolution of mathematical knowledge, but the various constraints shape mathematics, restrain it, and impose on it some uniquely determined paths.

The second narrative axis surveyed the issue of mathematics' foundational epistemological and ontological position, as presented by the paradigm of Kant's synthetic a priori. Does mathematics express an independent and formative source of knowledge, or is it reducible to other epistemological and ontological foundations?

The account of mathematics in this chapter highlights some of its distinctive features, and the notion of analytic a posteriori does express a claim to an autonomous and formative position, even though this position reacts to all sorts of a posteriori circumstances. Moreover, when viewed in the wider context of scientific and nonscientific forms of knowledge, it is clear that mathematics occupies a distinct position on a continuum of institutions of constrained knowledge, captured, for example, by its consensus-forming arbitration mechanism.

Still, evaluations of the position of mathematics depend more, I believe, on acknowledging the many different kinds of constraints imposed on mathematics, including those that are comparable to the constraints imposed on other knowledge institutions, rather than on highlighting only those constraints that make it appear exceptional. The underlying issue is not that of reducing mathematics to something else, but that of acknowledging the constellation of forces that constrain mathematics. At the same time, we should acknowledge that there's no *single* ground that bestows on mathematics a distinguished fundamental position.

The third narrative axis raised the issue of barring, taming, or living with monsters—those mathematical entities that impose inconsistencies or force the mathematician into conflict with established views. What does the preceding discussion tell us about handling such monsters?

The main point I wanted to make in this context is that even if we can bar or tame monsters by formal means, that is, by regimenting mathematical languages or delegating the power of arbitration to formal languages, monsters will continue to permeate mathematics. These monsters live at the level of interpretation. It's not that we have to hold on to all relevant interpretations all of the time. But our reconstruction, superposition, and deferral of interpretations are monstrous in their own way.

In contemporary mathematics, when these interpretation strategies generate conflicts and mathematical monsters, we can count on the arbitration of formal languages to keep mathematical claims in line. In the past, the latter resolution of ambiguity did not apply in quite this way, and led to some substantial debates over the legitimacy of some well-reasoned mathematical entities and claims. But either way, mathematicians did and still do use ambiguities productively. Monsters are a mathematician's friends.

Finally, there's the narrative axis of authority—the complex coalitions and antagonisms between mathematicians, philosophers, scientists, and figures of authority. How should we read these changing configurations of discursive alliances and conflicts?

Contemporary analytic philosophy tends to reflect on mathematics from the point of view of empirical sciences (or at least of the philosophy of empirical science). In analytic philosophy, there's a certain tendency to demure: how can a belligerent discipline such as philosophy claim to criticize such a successful venture as science? It seems that philosophers of mathematics see their authority as dependent on their support of science, rather than their critique.

This is not a very new trend. Wittgenstein asserted that "it is no good my getting the rest to agree to something that Turing would not agree to" (Wittgenstein 1976, 67–68; Turing being the only mathematician in the classroom). Even Latour, one of those "fashionable figures in the history and sociology and anthropology of science" denounced by Burgess and Rosen (1997, 241), believes that the objections of scientists to his anthropological account do matter, and therefore tried to rearticulate them in ways that would be more acceptable to scientists (Latour 2013, 12–13).

My objective in the last sections of this chapter was to rearticulate this focus, and phrase the questions concerning the truth and existence of mathematical statements and objects in a less restricted context, that of institutions of knowledge that do not necessarily belong to the empirical sciences. In that context, it becomes less obvious that the constraints imposed on mathematics necessarily align it with science.

Analyses of the position and merit of mathematics should evaluate the entire network of constraints imposed on this discipline, and reconsider its automatic alliance with natural science on the one hand, and analytic philosophy on the other. Such evaluation, I hope, will reconfigure the understanding of mathematics from the point of view of the humanities and the social sciences, and therefore help advance the stated goal of this book.

CHAPTER 4

■ ■ ■ ■ ■ ■ ■ ■

Two Case Studies of Semiosis in Mathematics

THIS CHAPTER ANALYZES TWO case studies of mathematical practice with signs. They highlight some of the aspects discussed in the previous chapter, especially around issues of interpretation.

The first case study follows the paradigmatic mathematical sign, x, as it is used in applications of powers series to combinatorics via generating functions. I try to show how x works across different formal contexts without having to account formally, either in real time or sometimes even post hoc, for its cross-context mobility. I try to show that the ambiguity and openness to reinterpretation of the sign make it a useful algebraic tool for solving an ever evolving host of problems.

The second case study concerns the use of gender representation in a family of combinatorial problems subsumed under the title of "stable marriage." The study of gender representations in scientific texts now has a long and distinguished history (for example, Cohn 1987; Keller 1992; Martin 1991; Potter 2001). I follow this tradition in showing how gender role stereotypes are reflected by mathematical discourse and constrain mathematical production. I further use this case study as an example of the vicissitudes of signs that cross between everyday and scientific contexts, and insist that while their trajectory tends to follow several known paths, their potential is unbounded and can't be inscribed in advance.

Both case studies try to substantiate my general claims about interpretation, formalization, and constraints over mathematical objects and statements. The first case study follows Wagner (2009c), and the second follows Wagner (2009b). For the benefit of accessibility, I removed almost the entire structural and post-structural theoretic framework that was elaborated in those papers. Readers who are interested in getting a better idea of my theoretical motivations and goals are welcome to check out the original papers.

Ambiguous Variables in Generating Functions

Between Formal Interpretations

When we think of a mathematical sign, we usually think of it as having a single fixed formal definition. If we fail to define the sign uniquely in a given context, we can't affirm the consistency of our work, and without consistency mathematics seems to make no sense. This is a prevalent view concerning mathematics and one of the characteristics that supposedly set it apart from other, less rigorous, forms of knowledge.

The following example qualifies this prejudice. It shows that we can actually work with a sign that is not subject to a single definition, and is interpreted in incoherent ways. We already saw (in the third section of chapter 1) that Wittgenstein believed that contradictions are not a problem as long as we don't pass through them. But this seemed to Turing like dangerous hand-waving. Here we see how this works in practice. In turn, this serves to show how we can mitigate the constraint of formal consistency in mathematical practice. Consistency turns out to be one constraint among many, not an absolute foundation.

This section considers the following kind of series:

$$\sum_{n=0}^{\infty} a_n x^n.$$

The context is that of solving combinatorial problems with so-called generating functions. In generating functions, the coefficients are usually the sequence of solutions of some combinatorial problem. Here the coefficients a_n are numbers. But when considering x, there are at least three preliminary interpretations.

First, x can be interpreted as a variable ranging over values in a given set. This interpretation obviously raises the question: which given set? In the context to be presented in this chapter, x is usually a real or complex variable, but this is not all there is to it. Depending on the coefficients, x may sometimes have to be restricted to a specific set of values in order to guarantee convergence. For example, if $a_n = 1$ for all n, then convergence requires $|x| < 1$. Some sequences of coefficients do not allow convergence for any x other than 0.

Another possibility is to view x as a transcendental element in an extension of the domain of the coefficients. This means that we think of x as a new kind of number, which, when subjected to addition and multiplication, obeys standard algebraic rules (associativity, distributivity, and so on), and which does not obey any identity of the form $\sum_{n=0}^{N} a_n x^n = 0$, unless for all n, $a_n = 0$.

Note that under this interpretation x is not a variable, but a constant. The "constancy" of x, however, is somehow underdetermined. Structurally speaking, if I replace the sign x by y, I'll get the exact same effects. In that sense, x and y would be the same. But to say that they are identical is risky. Indeed, if we want to use two elements transcendental with respect to the domain and to each other (x does not solve any polynomial equation with coefficients that may involve y and vice versa), then x and y will obviously be distinct, although structurally exchangeable. So the line between x as an individuated constant and as something more general here is blurry.

The main constraint of this last interpretation is that it does not allow for infinite series, so we must assume that $a_n = 0$ for all n from a certain point on. To work with infinite sums of powers of x, and avoid convergence issues, we can use the algebra of *formal* power series. In this context, x is not a variable or a constant, but a placeholder. We can think of it informally as a comma separating the coefficients of the series (so $\sum_{n=0}^{\infty} a_n x^n$ is just another way or writing $(a_0, a_1, \ldots, a_n, \ldots)$), or, more formally, as some sort of a λ-operator that assigns values (the coefficients a_n) to the sequence of integers (the powers n).

This point of view allows us to define and control sums of power series, their products, some divisions, and even derivatives and integrals of power series. For example, under this interpretation, the derivative has nothing to do with limits and differentials; it is simply the operator that takes as input the series $\sum_{n=0}^{\infty} a_n x^n$ (which is equivalent to $(a_0, a_1, \ldots, a_n, \ldots)$ or $\lambda n.a_n$) and gives as output $\sum_{n=1}^{\infty} n a_n x^{n-1}$ (which is equivalent to $(a_1, 2a_2, \ldots, na_n)$ or $\lambda n.(n+1)a_{n+1}$).

But this interpretation must be complemented and revised if we are to justify other tools of the trade, such as exponents of power series, their infinite products, continued fractions, Laurent series (where there are infinitely many negative powers of x), and, most importantly, substitutions of x by numbers. Indeed, all these operations may involve infinite sums of coefficients, which require either convergence considerations or some alternative formal account.

What we have here is an algebraic sign, x, that condenses various semantic and syntactic roles. This condensation might be sorted out by carefully distinguishing each use of x. But this careful, formal sorting out is not reflected in mathematical practice. As one author writes, "one of the attractions of the subject, however, is how easily one can shift gears from thinking of [power series] as mere clotheslines for sequences [the formal variable approach] to regarding them as analytic functions of complex variables [the numerical variable approach]" (Wilf 1994, 167). Algebra operates by condensing not only sets of values into a single variable, but also a variety of syntactic roles into a single sign.

When using algebraic techniques to explore and solve combinatorial problems, the analysis is conducted on the seam line between the various interpretations of the generating function. It may therefore be useful to view x as a variable, a constant, or an element of a formal power series. As Wilf explained in the preceding quotation, mathematical practice with generating functions requires us to "shift gears." In order to use divergent series together with analytic techniques, and exploit the entire arsenal of power series technologies, one doesn't choose how to articulate x. One takes advantage of the way x condenses various values and syntactic roles.

I conducted a small sample survey of textbooks and research monographs in order to probe into contemporary mathematical practice in the context of generating functions. As expected, and as I did in my own classroom, most authors either ignore the issue of the changing role of x, or wave it away with a cursory mention of some theoretic apparatus. Such comments are sometimes demoted to parenthetic remarks or footnotes, sometimes appear to be rather confused, and usually do not quite cover the entire scope relevant to the surrounding text.

One author, for instance, explains that "x is a formal variable and is used simply as an indicator," and that therefore "there is no need to question whether the series converges" (Liu 1968, 25–26). Then, in a footnote, the author suggests working "with the understanding that the value of x is set to be 0"—an approach that would get us nowhere fast, as it renders all power series identical. In fact, this author's practice includes substitutions of numbers into x and real function differentiations, which are not covered by the formal variable or zero value approaches.

Another form of ambiguity is manifest in juxtaposing the statement "convergence is not necessary for the series … to be useful in obtaining various properties of [its] coefficients," and the qualification that in dealing with "a power series expansion which is obviously *divergent* whenever [the variable is nonzero], we find it convenient to use [the symbol ≈ rather than =] to indicate divergence" (Srivastava and Manocha 1984, 78).

This nonrigorous notation (When exactly does divergence become "obvious"? What constraints do we place on its use?) is already available in MacBride (1971, 2), but neither text states the conditions under which their exponentiations, derivations, Laurent series and substitutions can be controlled under a formal series approach. To back things up, the latter author cites the work of E. T. Bell (as does Eisen 1969, 70), who qualifies the unknown in a footnote as "an abstract indeterminate," ignoring the real value limits and substitutions used shortly thereafter. Bell's work (1923)—a fine example of struggling for formal rigor in a setting where the means had not yet fully crystallized—does not quite provide the required theory of abstract indeterminates, and refers part of what is sought to Grassmann and Gibbs.

Other authors are somewhat more careful. One author provides an elementary theory of formal power series, explicitly marks transitions such as "below we treat the formal variable s as the *complex* variable $s \in \mathbb{C}$" (Lando 2003, 48), and raises some issues concerning continued fractions. But the same author does not question the consistency of switching between the formal and analytic approaches. A somewhat less rigorous approach is to make sporadic parenthetic remarks concerning radii of convergence, without following this concern throughout, as in Tucker (2002, 248). This approach is the successor of the most stoic approach of all: keeping entirely quiet about formal rigor, as practiced by the last prominent proponent of the nineteenth-century British school of algebra, Percy MacMahon (1915–16).

In this respect, Wilf is the most explicit author I found. He introduces both the formal and analytic theories, stating that under the formal interpretation "we love $[x]$ only for its role as a clothesline on which our sequence is hanging out to dry" (Wilf 1994, 6). He points out the technical transition between interpretations (Wilf 1994, 35), and even confesses a relationship of cheating, guilt, and making peace with his double-faced approach (Wilf 1994, 35, 30, 35, respectively). But his way of making peace is only almost convincing, and lacks a formal reconciliation especially around infinite products (Wilf 1994, 91). Wilf is also the author who makes ambiguity fully explicit: "One of the attractions of the subject, however, is how easily one can shift gears" (Wilf 1994, 32).

Before I bring on the wrath of mathematical authors, I must hasten to clarify that I fully endorse most approaches taken by the preceding authors. The theory isn't faulty. Generating functions are mathematically robust tools. Indeed, given sufficient context, any power series manipulations by the authors cited earlier *can* be justified rigorously. But my point is that the justification is not an a priori framework; the justification is a cohort of ad hoc constructions and shifting points of view that confront a developing set of tools. One can construct a theory unifying various formal power series and real or complex variable techniques, but then one could introduce new manipulations that exceed the constructed theory, and would require an extension and a review of the theoretic validation.

To understand contemporary mathematical practice, it is crucial to observe that the mathematical literature does not bother with such tailor-made theories. If they bother considering the issue at all, authors are typically happy with folk knowledge, based on experience and authority, that everything they do is rigorously justifi*able*. This justifi*ability* does not translate into an actual formal reconstruction or practiced conceptual stability, and *there's nothing wrong with that*. Logician Yehuda Rav states that

> Indeed, a strict axiomatization of analysis, or any field of mainstream mathematics, for that matter—certain geometries excepted—would be counterproductive, and essentially not feasible. The reason is simple, stemming from what can be called the transfer of technologies. The issue is this. Ideas

and concepts that were developed independently in a particular area of mathematics are frequently used and applied in a different context, much to the enrichment of the latter. Encapsulating a major branch of mathematics, such as analysis, for example, within a rigid axiomatic framework, would block beforehand such unforeseeable developments, including re-proving theorems by different methods, as mathematicians often do. (Rav 2008, 134–35)

Moreover, when one reaches theoretic forefronts, one encounters manipulations that produce correct results, but don't (yet?) have general rigorous mathematical frameworks to justify them, such as the *replica trick* from statistical physicists, which, in a sense, takes an integer valued variable, and then makes it converge to zero (see for example, Mezard et al. 1987 and Talagrand 2003).

The point I am making is that the sign x allows condensing various semantic and syntactic roles, a condensation that enables practices belonging to different justification frameworks to be performed on the same mathematical term without requiring an integrative a priori framework. Some ambiguity of roles is constitutive of working with unknowns in the context of power series and generating functions. Our instances of x earlier are not quite variables, constants, or λ-operators. The symbol x plays the role of a middle term that binds the system together.

The fact that it's possible, in principle, to formally reconstruct given portions of current mathematical practice does not make the ambiguities of really existing mathematical practice go away. Ambiguity does not mean that mathematical practice is faulty. The fault is with formal reconstruction, which fails to capture how mathematics is practiced. The single sign x retraces something that relates value-carrying variables, transcendental constants, and place-holding λ-operators—a practical relation that is independent of the articulations of the various syntactic roles.

Foundational philosophies of mathematics do not acknowledge indefinite deferral of formalization. For the formalist, logicist, intuitionist, platonist, empiricist, or structuralist, mathematical manipulations take place *in* distinctly and clearly articulated realms of rules, objects, or mental constructions, rather than *through* fluid interpretive frameworks. These philosophies may acknowledge sequential "snapshots" of mathematics, where each snapshot records a somewhat different robust mathematical system (Greek geometry versus Cartesian algebraized geometry; formal power series versus analytic power series), but these philosophies are not concerned with understanding the motion excluded from these snapshots. Indeed why should they? They are, precisely, foundational, and do not care for mathematical practice.

Models and Applications

Let's follow the use of x more explicitly by anchoring it to a specific mathematical practice: solving combinatorial problems with generating functions. I will not present the method in general, only supply an example.

Suppose we want to count all ways of distributing 20 identical balls into three boxes, where the first box may contain any number of balls, the second may contain up to 3 balls, and the third must contain an even number of balls. For example, 3,1,16 is one acceptable distribution of 20 balls into the three boxes, because 1 is not greater than 3, and 16 is even. 10,2,8 is another acceptable distribution of 20 balls. The problem demands that we count all such possible distributions.

It turns out that the number of acceptable distributions equals the coefficient of x^{20} after unpacking and simplifying the product

$$\left(x^0 + x^1 + x^2 + \cdots\right) \cdot \left(x^0 + x^1 + x^2 + x^3\right) \cdot \left(x^0 + x^2 + x^4 + \cdots\right)$$

(the powers inside each pair of parentheses reflect the restriction on the content of the corresponding box). This is the generating function of the preceding combinatorial problem.

This solution is based on an analogy between ways of distributing balls into boxes and ways of generating monomials (terms of the form x^n) when unpacking the product. To each legitimate distribution of 20 balls (say, 3,1,16 or 10,2,8), there corresponds a single product of monomials from the preceding product of sums that equals the monomial x^{20} (for example, the products $x^3 \cdot x^1 \cdot x^{16}$ or $x^{10} \cdot x^2 \cdot x^8$). Since there are as many products equal to the monomial x^{20} as admissible distributions of 20 balls, the coefficient in the sum of these monomials (which we get by unpacking the preceding product of sums) is precisely the required number of admissible distributions. (This is an easy observation, but somewhat tricky to explain. If you don't get it, think of the simpler product $(x^0 + x^1) \cdot (x^0 + x^1) \cdot (x^0 + x^1) = x^0 + 3x^1 + 3x^2 + x^3$). There are three ways to get x^2 here: $x^0 \cdot x^1 \cdot x^1$, $x^1 \cdot x^0 \cdot x^1$ and $x^1 \cdot x^1 \cdot x^0$; they sum up to exactly $3x^2$.)

Let's state the elements of the analogy between the combinatorial problem and the algebraic product that underlie the solution: the three pairs of parentheses in the algebraic model stand for the three boxes; the powers inside each pair of parentheses stand for the number of balls allowed in each box; the power 20 in the target monomial stands for the total number of balls; the products between the sums stand for a conjunctive relation between the conditions set on the boxes (at most three in the second *and* an even number in the third); the summation of monomials inside parentheses stands for a disjunctive relation between the possibilities for each box (either zero *or* one *or* two *or* ...).

But having correlated the elements of the problem and the elements of the solution so neatly, there's something that is conspicuously left out. Indeed, what does x stand for? Well, in fact, x here does not stand *for*. The values that may be taken by x have nothing to do with the problem. x is a term *in* an algebraic structure that binds the various algebraic elements in ways analogous to the combinatorial problem. x is a means to tie together an algebraic structure and a combinatorial problem. x is a signifier without a signified, the excess of the algebraic over the combinatorial. x is the degree of freedom of algebra that allows it to adapt itself to the combinatorial problem.

This algebraic apparatus has been adapted for solving a long string of changing problems throughout its history. How is it that a single language can adapt itself to a changing world of reference? According to Lévi-Strauss (1987, 64) every signifying order (in our case, algebra) requires "zero symbolic value" elements, such as our x, so as to guarantee the flexibility required for successful application to a changing signified reality (in our case, combinatorics).

The algebraic unknown is useful because it does not require us to decide once and for all what it stands for. We can, of course, stop at any given moment, and reconstruct a more-or-less stable signified for the algebraic unknown. But "even though scientific knowledge is capable, if not of staunching it, at least of controlling it partially [the] *floating signifier* [remains] the disability of all finite thought" (Lévi-Strauss 1987, 63). It exceeds finite thought in being reconstructible and reinterpretable, never reducible to a finitely containable carrier of an a priori order.

The algebraic unknown is the felicitous "expression of a semantic function, whose role is to operate despite the contradiction inherent in it" (Lévi-Strauss 1987, 63)—the contradiction between the formalized order of signifier (algebra) and the *changing* problems that we try to solve with it. It's the leeway of reworking signifier/signified relations through a "floating signifier" or "zero symbolic value" that enables us to more or less successfully represent combinatorics by algebra despite their discrepancies. The "floating signifier" is not a pathology, it is a constitutive condition for signification.

Openness to Interpretation

Let's return to our case study. By the rules for summing geometric series, the preceding product of sums equals

$$\frac{1}{1-x} \cdot \frac{1-x^4}{1-x} \cdot \frac{1}{1-x^2}.$$

When working with generating functions in this kind of rational function representation (products and fractions of polynomials), solutions sometimes go through other algebraic problems, such as turning products into sums. For instance, we may need to find the coefficients a and b for which

$$\frac{1}{(1-x)(1-2x)} = \frac{a}{(1-x)} + \frac{b}{(1-2x)}$$

or, equivalently,

$$1 = a(1-2x) + b(1-x)$$

Here, one of the simplest ways to get a and b is to substitute the values 1 and $\frac{1}{2}$ for x. This yields $1 = a(-1) + 0$ and $1 = 0 + b\frac{1}{2}$, respectively, so $a = -1$ and $b = 2$. Note that this requires the unknown x to become a substitutable variable, even if so far we have used it as a term of formal series. More importantly, note that x is substituted by the very values that render the first equality undefined!

As with the zigzagging between formal series and analytic functions earlier, this glitch too can be rigorously circumvented. For example, we could substitute values tending to 1 and $\frac{1}{2}$ from below, and take the limit—which, by the way, would still be problematic if we went back from the rational function to the power series, one of which is undefined around 1 ...

But such reconstruction is precisely what mathematicians usually do not do, even when teaching or writing for first-year students, who are not likely to note the gap and patch it up by themselves. Which, I emphasize again, is a practice that I endorse. Indeed, if this weren't our practice, how would our students ever get to actually practicing mathematics? One doesn't train new mathematicians by drowning them in mathematical foundations.

The point here is not just to demonstrate the superposition of various given approaches, or the floating of algebraic unknowns over an ever changing combinatorial signified. The point here is to bring up an openness to something that's not confined to a given, closed set of tools.

This openness has several dimensions. The first dimension is that of unexpected interpretations of algebraic unknowns. For example, suppose our manipulation would have led us to seeking a and b for which

$$x^2 - 1 = a(1 - x) + b(1 - x)^2.$$

We can try to use the same technique as earlier, but here no number substituted for x could nullify the b term without nullifying the a term as well. However, we can find matrices that do the job, such as

$$\begin{pmatrix} 1 & -1 \\ 0 & 1 \end{pmatrix}.$$

But if we allow matrices into this practice, the range of x will exceed the previous confines of commutative algebra without zero divisors that reigned over the discussion so far, and the entire theoretical justification would have to be reconsidered.

This first dimension of openness, the one articulated in this example, is the openness of the algebraic unknown to assuming new roles and reforming the justification framework of its practice. (Note also that if we allow x to take matrix values, then the number 1 becomes the identity matrix I, so should we still think of this 1 as a constant?)

The next dimension of openness is more radical than the earlier referential and syntactic openness. This second dimension of openness is the mathematician's practice of not deciding, either in advance or after the fact, what x is. This openness is the openness to semirigorous entities in mathematical practice.

How does that fit with my portrayal of formalization as the supreme arbitrator over mathematical validity? Since formalization plays the role of highest arbitrator, it is only called upon where there's a dispute that survives "lower" instances of arbitration, such as authority, experience, intuition, analogies, and standard tool boxes. What's settled on those grounds does not require formal arbitration. And even when formal arbitration is called for, it involves only partial formalization of subarguments, because full-scale formalization is unmanageable for real-life advanced mathematical disputes. Deeper levels of formalization may simply not be called upon, unless a specific challenge arises.

In the last example, x is not quite an element of the formal algebra of power series, nor is it a variable standing for values. It is not simply a middle term either. There, x is something that exceeds rigorous formality. Indeed, x can (to an extent, even should) be reintegrated through a formal reconstruction of the algebraic setting, which would render consistent the substitution of x. I have no wish to deny that such reintegration is an aspect of contemporary mathematical practice. But I insist on asserting that deferring rigor, sometimes indefinitely, is no less important in its turn.

On top of being a part of some formal language, x is a sign that's practiced in various material ways, all having to do with Derrida's condition of iterability—being quotable and transportable between contexts. These practices are not suppressed or replaced by formal and rigorous aspects of mathematics. These practices live alongside formalized practices, and mathematics is the joint fruit of the constraints and opportunities imposed by formal and informal aspects of mathematical practice superposed.

Gendered Signs in a Combinatorial Problem

The Problem

Let's begin this discussion with two compact formal statements of the problem addressed by Hall's theorem (Hall 1935), as presented in a more recent textbook (Wilson 1996, 112; the technical terminology will be explained shortly):

> If there is a finite set of girls, each of whom knows several boys, under what conditions can all the girls marry the boys in such a way that each girl marries a boy she knows?

And here is how the same problem appears on the next page:

> If $G = G(V_1, V_2)$ is a bipartite graph, when does there exist a complete matching from V_1 to V_2 in G?

The latter formulation is a technical question about "graphs," which are simply configurations of points and of lines that connect some pairs of the points. This question dates back to the 1930s (or even earlier, according to some authors). The gendered formulation, which very often accompanies the more abstract one, is attributed to Hermann Weyl (1949). In this formulation the points in the configuration become boys (elements of the set V_1) and girls (elements of the set V_2), and the lines connecting boys and girls become relations of acquaintance. The *marriage problem* is to find whether there exists a *complete matching*—a monogamous heterosexual marriage arrangement, which includes all girls, and which matches each girl to a boy she knows.

Our task in this chapter is to outline the interrelations between mathematical practice and gender language in the context of variants of this problem, and to explore their mutual impact. In other words, the task is to explore the constraints and opportunities that the gender language imposes on the practice of mathematicians in this specific context.

We will focus on a more recent version of the problem, introduced in Gale and Shapley (1962), which includes a component of personal preferences. Let us review this version in more detail before we survey the use of gender roles in various textbooks and papers studying this problem over the last four decades. Our initial presentation will be borrowed from one of the most important texts presenting the problem (Knuth 1997, 1; originally published in French in 1976).

> Let H and F be two finite sets of n elements. H is *the set of men* $A_1, A_2, \ldots, A_n$
> and F is *the set of women* $a_1, a_2, \ldots, a_n$. A *matching* is ... a set of *n* monogamous marriages between the men and the women.... Suppose that each man has an order of preference for the women and each woman an order of preference for the men.... A matching is *unstable* if a man *A* and a woman *a*, not married to each other, mutually prefer each other to their spouses. This "liaison dangereuse" occurs when:
>
> * *A* is married to *b*;
> * *a* is married to *B*;
> * *A* prefers *a* to *b*;
> * *a* prefers *A* to *B*.
>
> (The opinions of *b* and *B* are irrelevant here.) A matching is *stable* if this situation does not occur.

Here's an example from Steven Rudich's classroom presentation (2003). It includes a diagram (figure 4.1) representing five boys and five girls together with their ranking of members of the opposite sex, and an instance of a stable matching (here "pairing").

You can verify that any girl and boy not married to each other (for example, boy 1 and girl 2) do not both prefer each other over their actual spouses. Indeed, while boy 1 does prefer girl 2 over his actual wife (girl 5), girl 2 prefers her assigned husband (boy 2) over boy 1. Therefore, boy 1 and girl 2 do not form a potential "rogue couple," and according to the rules of the problem they will not break their marriages in order to marry each other. You can go on verifying that in the preceding example there is indeed no rogue couple.

It is not clear at first sight whether all systems of preferences admit a stable matching. Gale and Shapley proved that such a matching does indeed always exist, and supplied a simple algorithm (or set of instructions) for producing a stable matching. Before commenting on their algorithm, it is important to emphasize that this algorithm is *not* the only method for producing a stable matching. It is, however, historically the first to have appeared, and the most widely reproduced in the literature (although, I believe, not the simplest available).

STABLE PAIRINGS

A pairing of boys and girls is called stable if it contains no rogue couples

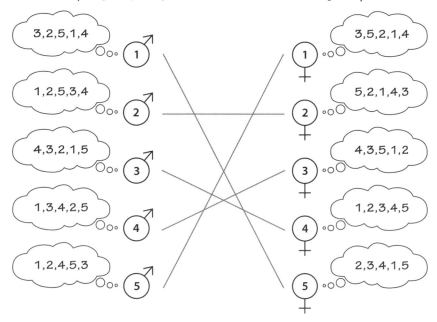

Figure 4.1: An example of a stable matching. Reprinted from Steven Rudich, "The Mathematics of 1950s Dating: Who Wins the Battle of the Sexes?" Available at http:/ www.cs.cmu.edu/afs/cs.cmu.edu/academic/class/15251/discretemath/Lectures /dating.ppt.

A compact description of the process of generating a stable matching can be found in Bollobás (1988, 86), where it is referred to as "simply the codification of the rules of old-fashioned etiquette: every boy proposes to his highest preference and every girl refuses all but her best proposal," keeping her favorite suitor on hold. Each rejected boy continues to propose to his next highest preferences, and each girl continues refusing all but her highest preference among the boys who actually propose to her at any given time, possibly rejecting a boy whose proposal she had previously kept on hold. "This goes on until no changes [new proposals] occur; then every girl marries her only proposer she has not yet refused."

TRADITIONAL MARRIAGE ALGORITHM

For each day that some boy gets a "No" do:
- **Morning**
  - Each girl stands on her balcony
  - Each boy proposes under the balcony of the best girl whom he has not yet crossed off

- **Afternoon (for those girls with at least one suitor)**
  - To today's best suitor: "Maybe, come back tomorrow"
  - To any others: "No, I will never marry you"

- **Evening**
  - Any rejected boy crosses the girl off his list

If no boys get a "No":
Each girl marries the boy to whom she just said "maybe"

TRADITIONAL MARRIAGE ALGORITHM

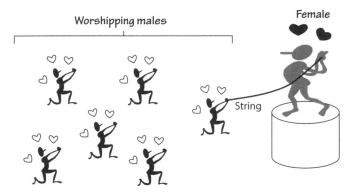

Figures 4.2 and 4.3: Literal and figural presentations of the Gale-Shapley algorithm. Reprinted from Steven Rudich, "The Mathematics of 1950s Dating: Who Wins the Battle of the Sexes?" Available at http://www.cs.cmu.edu/afs/cs.cmu.edu/academic /class/15251/discretemath/Lectures/dating.ppt.

A more graphic description (figures 4.2 and 4.3) of this process is available, again, from Rudich (2003). Gale and Shapley proved that this iterative process must converge to a one-to-one matching in a finite number of steps, and that this matching is indeed stable in the sense defined earlier (no rogue couples).

Gender Role Stereotypes and Mathematical Results

It is no surprise that mathematicians, like almost anyone else in our society, reproduce gender role stereotypes. In this specific setting, the reproduction of stereotypes can be somewhat excused by noting that it is precisely the stereotypes that make the marriage terminology an intuitively accessible presentation of the problem. If stereotypes were challenged in the presentation, then practitioners wouldn't be able to build on their intuitively accessible stereotypes in order to understand the question, and the entire imagery would be useless. But, as we shall see, adhering to the stereotypes has its price. I emphasize that my point is not to accuse authors of sexism. My point is to analyze the constraints and opportunities that the use of stereotypes (here, gendered stereotypes) in presenting such mathematical problems may impose.

In the dozen or so textbooks, monographs, and research papers that I surveyed for the purpose of this analysis, I found substantial variety concerning notation, terminology, and focus on men or women (most texts in fact focus more on what women do in the process of the matching algorithm). Rudich's slides, for instance, are exceptional in the way they put women, literally, on a pedestal (as we shall see later, this is a deliberately ironic choice). But as one would expect, throughout the literature it is the boys who propose.

A typical statement in this context is: "We will adopt the traditional approach describing the men as 'suitors' in a 'courtship' process, but analogous results may be obtained by reversing the roles of the sexes" (Gusfield and Irving, 1989, 8). The problem is symmetric—the roles of men and women may be interchanged—but the literature consistently opts to solicit the men to propose. The only exception that I found to this rule (excluding texts that do not quote the matrimonial imagery at all) is a text that does use matrimonial language, but refuses to assign genders: "We leave the reader to assign the sexes of V_1 and V_2." (Asratian et al. 1998: 67).

A particularly disturbing feature in the various narrations of the algorithm is that in *none* of the surveyed texts do the women ever say "yes" to the marriage proposals. Their replies are either a definite "no," a deferring "maybe," or a silence that is interpreted as provisional or permanent consent. At the end of the algorithm, no additional step is required for all the final silences and "maybes" to become consummated in marriages, so the women are not required to actively confirm. From the point of view of implementing the algorithm in computational contexts, this makes perfect sense; but the fact remains that a mathematical woman needn't say "yes." It suffices that she does not explicitly refuse.

The result of the Gale-Shapley algorithm is very strongly biased in favor of the proposing side (which is a euphemism for men). In fact, given a system of preferences, the Gale-Shapley algorithm produces the stable matching that is worst for all women and best for all men. This does not mean that each man gets his first choice and every woman her last; indeed, such a matching needn't be stable nor even monogamous. What "female pessimality" and "male optimality" mean is that no *stable* matching exists, where any man marries a woman whom he prefers over the one assigned by the Gale-Shapley algorithm; on the other hand, no *stable* matching exists that marries any woman to a man less desirable to her than the one assigned by the Gale-Shapley algorithm.

As we noted earlier, the stable marriage problem is symmetric (the sides can be exchanged without changing the problem). Gender roles discourse however is asymmetric, as is Gale and Shapley's algorithm and its result. Should we conjecture that a link exists between these two forms of asymmetry? Later I will present some evidence suggesting that gender role stereotypes reproduced in theoretical deliberations may affect the structure of mathematical results.

But before I do that, I should clarify that I don't pretend to offer a conclusive proof that the language we use causally determines scientific results. First, I see no methodological scheme that can convincingly separate scientific results from the language used to express them so as to assert a causal relation between distinct variables. Second, to the extent that we can articulate this separation, I doubt that we could come up with a corpus of texts that would supply proper control over the variables we might wish to correlate. Given these circumstances, my goal is to make room for the possibility that language co-constitutes scientific (specifically, mathematical) production. Accepting this very possibility, regardless of our capacity to provide a rigorous proof, has significant political and ethical repercussions: it means that scientists should assume responsibility for their language.

Let's review, then, the evidence that I propose for the possible impact of gender role discourse on the mathematical analysis of the stable marriage problem. The fact that the Gale-Shapley algorithm is male-optimal is noted in Gale and Shapley's original paper from 1962. It took, however, an additional nine years for the observation that the algorithm is female-pessimal to reach the literature (McVitie and Wilson 1971). Both facts are just as easy to observe and prove. Should we interpret this temporal gap as reflecting a foreclosure on stating the question from the women's point of view?

The Gale-Shapley algorithm is indeed strongly biased in favor of the proposers. Today, however, there are several algorithms for producing stable matchings that generate results of a more balanced nature. It is interesting to observe that the first steps in this direction were quoted by a subsequent author with no reference to a publication (Selkow, quoted in Knuth 1997, 51; Conway's median observation, quoted in Knuth 1997, 56). Should we interpret this deferral of publication as indicating a lack of interest in generating balanced algorithms?

The first kind of more balanced algorithms start with the Gale-Shapley algorithm, and manipulate it by exchanging spouses in controlled ways. In the presentation of these algorithms, the elements of analysis undergo various transformations. The basic unit of analysis corresponds to the symmetric notion of "possible pair" rather than to men or women as in the original algorithm (Gusfield and Irving 1989; Vande Vate 1989; Teo and Sethuraman 1998). Can the shift toward a symmetric unit of analysis account for the emergence of more balanced algorithms?

The second kind of more balanced algorithms emerges from Roth and Vande Vate (1990), further developed in Ma (1996) and in Romero-Medina (2005). The basic step in these versions is applying the Gale-Shapley algorithm in a way that allows all players to assume the proposing and reactive roles. The algorithm and its analysis are (in my opinion) simpler than Gale and Shapley's algorithm: people enter a room one at a time. Each new entrant is matched to her or his highest preference among all those in the room willing to accept the new entrant (either because they are single or because they prefer the new entrant over their current partner). Now, the matching of the new entrant may have separated someone in the room from a preexisting spouse. Such newly single persons are then matched as if they have just entered the room one by one. This goes on until no more couples in the room are broken. Then a new person enters the room, and so on, until everyone is matched.

This algorithm is practically identical to the Gale-Shapley algorithm, except that it conducts the proposal sessions individually, rather than in men/women blocks. Nevertheless, when it came into the theory (in 1990), this minor variation was quite a novelty—so much so that it solved an open problem raised by Knuth fourteen years earlier. Should we explain this long delay by an adherence to gender role stereotypes?

These pieces of circumstantial evidence can be confronted with various critiques. Obviously, one could offer alternative explanations to the theoretic developments that would have nothing to do with gender or language. I do not claim that such explanations are necessarily invalid. My objective is to open up the *possibility* that gender role discourse could have had substantial impact on the development of the theory. I will therefore only comment specifically on one possible objection.

One could claim that more balanced algorithms emerged later in the theory simply because such algorithms are more difficult to produce. For some—but not all—of the more balanced algorithms, this claim is indeed true. In fact, one version of the problem of finding a balanced stable matching, the so-called sex-equal matching problem, is known to be NP-hard (Kato 1993; NP-hardness implies that it is not practically computable in some sense).

However, this excess of difficulty could itself be an effect of language. We know, for instance, that geometric problems may change their degree of difficulty if presented in terms of calculus or classical geometry. Even the notion of NP-hardness itself depends on a contingent construct of computational complexity, which shouldn't monopolize our notion of hardness of computation. (I do not refer only to exotic models such as DNA and quantum computing, but also to the asymptoticity built into computation complexity definitions, which doesn't always reflect "real life" concerns.)

Before we move on with the argument, I would like to note a paper dealing with another combinatorial problem, the so-called ménage problem. The abstract reads:

> The *ménage problem* ... asks for the number of ways of seating n couples at a circular table, with men and women alternating, so that no one sits next to his or her partner. We present a straight-forward solution to this problem. What distinguishes our approach is that we do not seat the ladies first. (Bogart and Doyle 1986)

The authors claim, with a certain reservation, that "of all the ways in which sexism has held back the advance of mathematics," the fact that all previous algorithms seated the ladies first "may well be the most peculiar" (Bogart and Doyle 1986, 517). It seems that at least some mathematicians directly involved with such subject matter acknowledge the constraints imposed by gender role stereotypes on mathematical discovery, at least as a peculiarity.

Mathematical Language and Its Reality

It would be extremely uncharitable to assume that any of the cited researchers genuinely believes that the stable marriage model is a perfectly adequate description of social reality. Indeed, the gap between the model and the supposed reality is not ignored in the literature. One can read that "We will sometimes speak about courtship, but never of dependent children or mid-life crises" (Roth and Sotomayor 1990, 1), and that "This makes the (somewhat unrealistic) assumption that it is always better to be married (to an acquaintance) than to stay single" (Bollobás 1998, 85). Some texts consider the entire marriage terminology as "frivolous" but continue to use it anyway (Wilson 1996, 113).

Nevertheless, the literature does contain statements that indicate that the analysis and its results are taken as representative of reality, such as "Societal habits thus favor men" (West 1996, 177), and "We shall give a result which perhaps demonstrates an effect of complete inequality of the sexes" (Asratian et al. 1998, 70). When one reads that "The algorithm is the codification of old fashioned etiquette" (Bollobás 1998, 86), one reads a rather doubtful and revisionist social history.

The realistic interpretation of the mathematical problem is reinforced when shifting from the marriage context to the job-assignment formulation of the same problem. The latter is considered as "more serious," even though this version too abstracts and ideologizes many factors that drive mathematical analysis away from any social reality that it purports to seriously emulate.

The most widely discussed applications of stable matching schemes are the matching of residents to hospitals and of students to schools. If we consider information gaps (people don't always know each other's preferences, and so can't tell if there's an opportunity to "elope"), change of mind and hesitation (preferences are contingent, change over time, and may depend on what is learned over time of others' preferences), contractual obligations, and practical impediments (which may prevent "rogue couples" from "eloping" even if such couples do exist), stability may become less relevant.

The applicability of stability to actual, real life contexts has only been considered seriously since the mid-1980s. Roth and Sotomayor (1990) survey some aspects of stability in applications. To be fair, their analysis is lucid enough to raise many pertinent objections, including some of those mentioned earlier. They even write (1990, 156): "at least one of the authors would feel very differently about the theory presented here if the weight of the empirical evidence were different." But as is often the case in mathematical game theory and economy (recall our Black-Scholes vignette), the interpretation of empirical evidence allows ideological commitments to gain the upper hand. It was in fact the very evidence raised by Roth and Sotomayor that evinced my suspicion that the success or failure of a matching scheme in contemporary applications may not depend heavily on stability, but rather on other factors.

In fact, the Gale-Shapley algorithm was introduced in 1962. It had, however, already been in use for eleven years by that time. Since 1951, residents (formerly interns) have been assigned to US hospitals via a centralized application of a variant of the Gale-Shapley algorithm. That this was the case was observed by Roth only in the mid-1980s. Is it possible that the notion of "stability" is an a posteriori justification for the coincidence between a biased assignment scheme and a biased mathematical discourse, both emerging from interrelated ideological commitments? To quote Emily Martin quoting David Harvey, perhaps this amounts to the "implanting of social imagery on representations of nature so as to lay a firm basis for reimporting exactly that same imagery as natural explanations of social phenomena" (Martin 1991, 500).

I will not ponder here over the details of the debate. I would only like to indicate that the quest for stable solutions is an overarching characteristic of contemporary game-theory and economics. Many questions related to the production of "mobility"—which is often marked as a desirable social goal—are foreclosed by some aspects of contemporary game-theoretic and economic language.

In our context, this foreclosure is expressed through statements such as: "A pairing is doomed if it contains a rogue couple" (Rudich, 2003); "The idea of stable matching is inspired by the search for an idyllic society of married couples" (Asratian et al. 1998, 67); and "from the practical and algorithmic standpoint [some specific] results are negative, since they do not lead to acceptable ways to produce desirable stable matchings using the Gale-Shapley algorithm" (Gusfield and Irving 1989, xv). Another strong statement is the following:

> One of the main difficulties with using the men-optimal stable matching mechanism is that it is manipulable by the women: some women can intentionally misrepresent their preferences to obtain a better stable partner. (Teo et al. 1999, 430)

Rather than describe preference list manipulation as a way of turning a highly unbalanced algorithm into a more balanced one, the literature condemns this opportunity as "cheating," namely, as something that should be prevented.

The Forking Paths of Mathematical Language

It may seem as if I were advocating the banishment of gendered language from mathematics in favor of some pure unbiased abstraction. But this is not the case. First, the supposed "purity" is unattainable. Our language is imbued with gender, whether we like it or not. Some theoreticians go as far as to say that every discourse that operates a binarism operates a gendered structure.

But the desire for a purified scientific language encounters a further, more radical problem. This desire to purify science from linguistically imposed bias relies on an assumption that behind the various ways of representing scientific or mathematical problems there is an objective and originally present problem—that our presentation of a problem covers over a "pure" problem. I would rather promote a stance according to which the scientific or mathematical problem is, precisely, the various ways in which it is re-presented and applied—that mathematical problems are not simply covered by the symbolic fabrics and practices into which they are sewn, but are actually spun from those very symbolic fabrics and practices, and consist of quilting together various symbolic and practical fabrics. There needn't be any essence to a mathematical problem other than an essential dependence on the texts and practices through which the problem is articulated.

As explained in the previous chapter, accepting such a stance does not imply that mathematics is arbitrary or unconstrained. It does mean, however, that scientific and mathematical authors are implicated in a responsibility for the ideological commitments and opportunities in which scientific production is immersed.

To the two objections mentioned earlier I would like to add one further objection. Entangling abstract mathematical patterns with some looser everyday linguistic structures might actually help mathematicians think creatively. There's even a small chance that this entanglement may help shift some of our everyday biases.

To substantiate this position, I will go through some examples of semiotic-mathematical transformations and their interaction with ideological commitments. As we shall see, this interaction is often conservative and mutually reinforcing, but it can also transform, manipulate, and undermine the ideological commitments of the discourses with which mathematical production interacts. The interaction between mathematics and its signs cannot be reduced to a straightforward, conservative, and reactionary imposition.

The variations of the marriage problems include what we might call "alternative families." For example, instead of a matching based on a gender division, we may be given a set of people each of whom can be matched to any other. The literature I surveyed never refers to this as the "homosexual marriage problem." Rather, it is called the "roommates problem," and even "a unisex problem: stability of roommates" (Knuth 1997, 64). The one exception is Rudich's presentation, which describes this situation as "bisexual dating." The role of this scenario in Rudich's presentation, however, is to demonstrate a case where a stable matching is unattainable. Since stability is strongly marked as an essential ideal in Rudich's presentation ("a pairing is doomed if it contains a rogue couple"), bisexual dating seems to be negatively marked.

The preceding example demonstrates how, when the discourse encounters an opportunity to endorse progressive aspects of sexuality, it tries to clean itself up and revert to gender-neutral language. Conservative sexuality discourse is strong enough to hold back the application of sexual metaphors to mathematics in unorthodox situations, even when progressive sexual language makes mathematical sense.

The same phenomenon is repeated when other combinations are suggested. For instance when triples, rather than couples are to be matched, the three roles in most presentations are either men, women, and children or men, women, and dogs (for example, Knuth 1997, 64). This choice reiterates both normative family stereotypes and the position of women in such families, while foreclosing the horizon of polyamorous relationships.

Yet another instance for this form of conservatism is the construct theoretically referred to as "chains" or "rotations." This term refers to a set of couples, who exchange spouses according to a certain rule. The literature never describes this situation as "swinging" or "open relationships." Rather, at that point the texts opt to revert to a technical gender-free language.

But nonorthodox sexuality does manage to occasionally emerge in mathematical discourse, despite its conservative tendencies, as the following quote demonstrates (the context is an attempt to reduce one variant of the marriage problem to another): "Such a reduction might involve turning each person into two persons, one male and one female (splitting one's personality into its animus and anima)" (Gusfield and Irving 1989, 221). The splitting, which mathematically makes perfect sense, imposes some form of queerness on the underlying concept of gender.

One of the most important variations of the marriage problem deals with a situation where each man may marry several women. Some authors refer to this variant as the "harem problem" (for example, Wilson 1996, 114). Most authors, however, shun the use of marriage terminology for this variant. "It is convenient to use politically correct terms," writes Bollobás (1998, 90), and turns to the student-college formulation of the problem, where each college should be matched to several students.

But this attempt to clean up the mathematical discourse from reactionary gender imagery ends up generating progressive opportunities for cross-gendering. In translating the man-woman matching problem to a student-school matching problem, it is not clear which should play whose role. Thus, Teo et al. (1999) assign the women to the role of students, whereas Immorlica and Mahdian (2005) make the opposite choice. A person reading both papers, and trying to retain the contents of both papers at once, is confronted with the necessity to allow a simultaneous bivalent role allocation for men and women. This is a much more intricate endeavor to sustain than simply choosing between the mainstream allocation of gender roles and its inversion.

On some occasions, the gender imagery is confronted with other images, which may or may not reinforce gender role stereotypes. Thus, when Knuth (1997) demonstrates links between the stable marriage problem and other algorithmic problems (related to path finding and hash tables), the transformations rely on such statements as "The city ... plays the role of the woman" and "The cell that each [man] occupies corresponds to the number of his girlfriend" (Knuth 1997, 35, 40). The image of the woman as a place that confines the man correlates with the ideological link between women and domesticity or women as ball-and-chain.

But such interactions between different problems do not necessarily work out so elegantly oppressive. Roth and Sotomayor (1990) regard the problem from an economic point of view and introduce the term "marriage market. " This commodification of the matrimonial situation is reflected in another paper by such statements as "The cost for man i to be married with woman j is $x(i, j)$ and the cost for woman j to be married with man i is $y(j, i)$" (Dzierzawa and Oméro 2000, 322). Then, following its own statistical physics inspired logic, the paper goes on to transform cost to energy, as men and women become species of particles. The effects of this semiotic drift in terms of gender representations cannot be anticipated in advance.

As a somewhat optimistic final statement, I will include the following quotation: "The man optimal matching corresponds to the minimal P-set.... Mathematically there is no problem with this, but the reader may have to make a psychological transformation to avoid unconsciously identifying (male) dominance with maximality" (Gusfield and Irving 1989, 75). Regardless of what a P-set is, this demonstrates how a perfectly viable mathematical transformation might impose a psychological reevaluation of ideologically tainted language.

But how seriously should we take these few and feeble mathematical lines of flight from gender role stereotypes? There is something a little too easy about such transformations. The fact that it is so easy for Knuth to write "It suffices to exchange the role of the men and the women and apply the theorem" (Knuth 1997, 57) stands in sharp contrast to the effort involved in actual change of extra-mathematical discourse and practice.

We could claim that the preceding effects of ideological replication and semiotic drift are confined to mathematical texts, and therefore have nothing to do with social change. We could then infer that we needn't bother at all with the languages that science uses, because these languages are restricted to a confined textual or discursive "reservation." But anecdotes of classroom discussions of the marriage problem, which quickly deteriorated to an opportunity for venting the slut-shaming tendencies embedded in the male-dominated environment of an undergraduate science course, may suggest otherwise. If we believe that there is some (however small) interaction between mathematical language and social reality (if only because it is spoken by real, at least somewhat socially functional people, who often teach students who don't necessarily end up being mathematicians), then the semiotic drift imposed on gender role stereotypes by mathematical transformations should not be ignored.

Perhaps, then, we should encourage mathematicians to explore conceptions that are feminist or queer. Perhaps we should encourage social and exact scientists to carry their latent and explicit ideological commitments through mathematics' obscure transformations. Perhaps this would lead us to explore new semiotic possibilities for confronting the impossible impasses in our ways of speaking gender and/or science. Perhaps encouraging signifiers to cut across discursive systems where they do not, supposedly, "belong," does have some therapeutic potential for our contemporary social malaise.

CHAPTER 5

Mathematics and Cognition

So far, we have looked at the manifestation of mathematics on the surface of its practice. But is there something hidden beneath the surface?

These days, it is hard to talk about ideal essences that lurk beneath surfaces. But it's becoming more and more common to try to figure out how things work beneath the surfaces of our skulls. If we can observe constraints on mathematical thinking that are wired into our brains, then perhaps the Kantian synthetic a priori is vindicated, and perhaps our conceptual mathematical freedom is more limited than surface appearances lead us to believe. If there are cognitive constraints on mathematical activity, perhaps they supersede other, external constraints on mathematical production.

This chapter will therefore engage some neuro-cognitive theories of mathematical concept formation. We will start with theories of elementary number processing (by Stanislas Dehaene and Vincent Walsh), and move on to more complex theories of mathematical concept formation (Lakoff and Núñez, Walter J. Freeman III). There is a substantial discrepancy between the former discussion of elementary number processing and the latter discussion of higher mathematics, but it seems that the division between modular and integrated approaches to cognition governs both debates. Therefore, a review of the former discussion does shed light on the latter, and it makes sense to tie the two debates together in this presentation. Finally, I tentatively explore a philosophical language, derived from Gilles Deleuze, which will try to express my take on things without presuming to be properly scientific.

The Number Sense

Our starting point is an articulation of arithmetic cognition associated primarily with Dehaene. By the term "number sense," Dehaene (2011, 227) means the following:

- The human baby is born with innate mechanisms for individuating objects and for extracting the numerosity of small sets.
- This "number sense" is also present in animals, and hence it is independent of language and has a long evolutionary history.
- In children, numerical estimation, comparison, counting, simple addition, and subtraction all emerge spontaneously without much explicit instruction.
- The inferior parietal region of both cerebral hemispheres hosts neural circuits dedicated to the mental manipulation of numerical quantities.

This number sense refers to such capacities as subitizing (detecting the cardinality of small collections), predicting the cardinality of a collection formed by combining two other small collections, and forming quantitative estimates for larger cardinalities. For example, rats can be trained to press a lever a certain number of times (with diminishing success as the number increases), monkeys can successfully estimate which tray with two separately presented piles of treats has a larger total number of treats, and very young babies show enhanced attention when two dolls separately go behind a screen, but only one is there when the screen is removed (compared to the attention given when the expected two dolls are revealed).

The neuro-cognitive problem is how these phenomena, as well as more advanced culturally acquired mathematical capacities, are implemented in the brain. Dehaene's answer is that these capacities most likely depend on neural circuits located in specific brain areas and dedicated to numbers. His claims are based on the usual tools of the trade: brain imaging of healthy people and observations of patients with localized brain lesions engaging in conscious and unconscious number-related experimental tasks.

Dehaene's model consists of three elements:

[A] quantity system (a nonverbal semantic representation of the size and distance relations between numbers, which may be category specific), a verbal system (where numerals are represented lexically, phonologically, and syntactically, much like any other type of word), and a visual system (in which numbers can be encoded as strings of Arabic numerals).... [W]e propose that three circuits coexist in the parietal lobe and capture most of the observed differences between arithmetic tasks: a bilateral intraparietal system associated with a core quantity system, a region of the left angular gyrus associated with verbal processing of numbers, and a posterior superior parietal system of spatial and nonspatial attention. (Dehaene et al. 2003, 488)

The first circuit provides "a nonverbal representation of numerical quantity, perhaps analogous to a spatial map or 'number line.'... This representation would underlie our intuition of what a given numerical size means, and of the proximity relations between numbers" (Dehaene et al. 2003, 489). The second "is part of the language system, and contributes to number processing only inasmuch as some arithmetic operations, such as multiplication, make particularly strong demands on a verbal coding of numbers" (Dehaene et al. 2003, 498). The third, "in addition to being involved in attention orienting in space, can also contribute to attentional selection on other mental dimensions that are analogous to space, such as time ... or number" (Dehaene et al. 2003, 498). This last system can deal with seriality or ordinality.

The three circuits show some unexpected forms of interaction. For example, when subjects are required to respond to number-related stimuli with their right or left hand (for example, press the right-hand key if a number stimulus is larger than 65 and the left-hand key if it is smaller—or vice versa), they tend to respond faster and better when the "larger" response is assigned to the right hand and "smaller" to the left. This is an example of the so-called SNARC effect (acronym for spatial-numerical association of response codes). It suggests that quantity is mentally associated with spatial position, linking "larger" with "rightward."

Despite these interactions, Dehaene tends to remain faithful to the hypothesis of a number-specific component in the brain. To explain the interactions between number responses and other dimensions (such as space or time), Dehaene argues the following:

Our brain imaging research has revealed that it comes from a systematic "leakage" of neural activity in the parietal lobe. When we evoke a mental representation of some numerical magnitude, brain activation starts in the hIPS [horizontal intraparietal sulcus], but also expands into nearby regions that code for location, size, and time. As a result, when we see a number, our space perception, and even our hand and eye movements, are affected by the slightly biased estimates that we make of these parameters.... I posit that recent human inventions [such as advanced arithmetic] have to find their niche in a human brain that did not evolve to accommodate them. They have had to squeeze themselves into the brain by invading cortical territories dedicated to closely related functions.... [T]hese novel [mathematical] concepts can only be represented in the brain, at least in part, because existing functions in the nearby cortex are recycled for this new use. Thus, arithmetic invades the nearby areas coding for space and eye movements. (Dehaene 2011, 245–46)

So by this view, we start with a number-specific circuit. Subsequently, human constructs and social training expand this number circuit, forcing it onto neighboring circuits devoted to spatial attention and symbol processing.

A dissenting view can be found in Vincent Walsh's ATOM (acronym for a theory of magnitude). The main tenets of this theory are (Walsh 2003, 484):

- Space, quantity and time are linked by a common metric for action.
- Time and quantity estimation operate on similar and partly shared accumulation principles.
- The inferior parietal cortex is the locus of a common magnitude system.
- The apparent specializations for time, space, and quantity develop from a single magnitude system operating from birth.

The main bone of contention with respect to Dehaene's number sense is whether number is an innate given that is later associated to space, time, and other quantities or a subsequent specialization that emerges from a generic sense for quantity. According to Dehaene, for example, the SNARC effect is the result of an interaction between initially independent number and space circuits. For Walsh, SNARC is a special case of SQUARC (acronym for spatial-quantity association of response codes), which derives from the fact that all kinds of quantities are processed by the same brain system (Walsh 2003, 487).

Walsh's position derives from a central tenet of embodied cognition: that our cognitive capacities reflect our embodied actions. Since we can't move in space without moving in time, and since counting is a temporal process most often applied to spatially arranged elements, space, time, and number should be originally interconnected.

> [S]haking hands, kissing, catching, throwing, playing an instrument, gathering kindling or paying by cash all require spatiotemporal coordination. Behavior may be spatial or temporal in a laboratory, but in the real world they originate in the same coordinate system applied to all magnitudes. (Bueti and Walsh 2009, 1836)

This fact of coordination between various kinds of magnitudes means that the brain does not initially distinguish between these dimensions. "In other words, the parietal cortex transformations that are often assumed to compute 'where' in space, really answer the questions 'how far, how fast, how much, how long and how many' with respect to action" (Walsh 2003, 486). It is only at a later stage, when learned concepts of number enter the field, that this action-magnitude brain component is appropriated to dissociate number concepts from the integrated action-magnitude system.

But the disagreement extends further than that. The question is not just whether number specification precedes or follows space/time/quantity associations. Scientists are also divided on whether we process numbers abstractly (that is, independently of how they are represented) by a pure number module, or via different brain mechanisms associated with different number representations (verbal, visual, and so on) that relate to an integrated action-magnitude system.

According to Walsh, when number concepts emerge from the action-magnitude system, they do not cohere to form a category-specific circuit independent of how numbers are represented. Instead, they form various components dedicated to specific number representations. These different representation-specific components build on the action-magnitude circuits without integrating into an abstract mental number representation.

If we prefer to put the question in functional terms, rather than in terms of neuro-physiology, then scientific interpretation is divided on whether "behavior depends only on the size of the numbers involved, [or] on the specific verbal or nonverbal means of denoting them" (Dehaene et al. 1998, 356). In computational terms, we may ask whether we perceive stimuli via representation-specific circuits but process them by means of a representation-independent circuit, or whether the processing is also carried out by representation-specific circuits.

Dehaene's challenge is to explain how, despite the distinct and purely numerical processing circuit, we manifest such strong relations between different expressions of magnitude (as in SNARC). Walsh's challenge is to explain how, despite the representation-specific processing that builds on a generic action-magnitude system, we manage to develop a unified and abstract understanding of number.

Dehaene meets his challenge by claiming that "the overlapping activation need not reflect a lack of neural specialization. Rather, analog quantities such as number, location, size, luminance, or time may well be coded by neuronal assemblies that are specialized, yet intermixed within the same voxels [strongly associated small sets of neurons]" (Dehaene 2009, 241). So the treatment of each kind of magnitude (most notably, abstract numbers) is distinct, but the neurons are so close together that they interact and can't be told apart by current imaging techniques.

Walsh, in turn, claims that "humans do not, as a default, represent numbers abstractly, but can adopt strategies that, in response to task configuration and demands, can create real or apparent abstraction." (Cohen Kadosh and Walsh 2003, 326). This stance follows Barsalou's tenet that "abstraction is simply a skill that supports goal achievement in a particular situation" (Barsalou 2003, 1184), pointing out that "dissociations between [various kinds of magnitudes] probably tell us more about the strange things we can make people do in experiments rather than how the brain operates in the real world" (Bueti and Walsh 2009, 1836). Abstract number is, in this view, a complex epiphenomenon, not something embedded in the brain.

At this point, the question becomes undecidable, at least given the limitations on experimental design and the flexibility of scientific interpretation. The confrontation of Cohen Kadosh and Walsh's argument for representation specific processing (2009) with the responses of various scientists that challenge their argument makes it clear that the current experimental data can, with some effort, support either position.

In fact, the question might not be very well posed. Suppose for example, that some neural microsystems are dedicated to numbers and process quantities independently of representation, while others process them in a manner that depends on specific forms of representation and generic action-magnitude systems; but suppose, crucially, that none of these circuits ever work, or can be made to work, without the others joining in and interacting with it. More specifically, suppose that what each of those microsystems does is not a whole operation (for example, take two numbers and yield their sum), but some preparatory subprocess (for example, break numbers into smaller numbers for further processing) that has to combine with abstract and non-abstract, dedicated and generic circuits to produce meaningful results. Or suppose that the number-dedicated abstract circuit always operates in conjunction with at least one of the representation-specific circuits (that build on some generic action-magnitude function), and that a correct result depends on a feedback loop between the two kinds of circuits. Suppose, therefore, that if you shut down either the abstract or the nonabstract, the dedicated or the generic processing circuits, then the whole systems fails to produce a meaningful response (even if we set aside the input/output issue). Under such assumptions, we would have some circuits doing abstract processing, but not a performative manifestation of an abstract number sense in real live humans. The notion of abstract processing would then be an artifact of neuro-computational analysis—but one that is nevertheless grounded in neuro-computational reality! (If I read them correctly, the response of Piazza and Izzard to Cohen Kadosh and Walsh [2009] may be pointing in this direction.) In such circumstances, deciding for or against abstraction or number dedicated processing is not a well posed problem.

My point here is that the very formulation of the problem seems to assume a commitment to conceptual distinctions that do not necessarily shape brain performance. Indeed, if we follow the superposing interactions explored in previous chapters (money and numbers, numerical and formal variables, graphs and weddings, and so on), we see that some seemingly clear conceptual distinctions cannot be neatly sorted out when applied to mathematical practice.

Before we bring up alternative scientific formulations of the problem of mathematical cognition, let's try to think of it in historical, rather than neuro-cognitive terms. One common way of telling the history of numbers is the following. First, there was pure Greek geometry. In this strand of geometry, lines did not have numerical lengths. Indeed, throughout Euclid's elements, no line segment is assigned a numerical length (the arithmetical books VII–IX do deal with numbers, but not as representations of geometrical entities). The only ways to relate geometrical magnitudes to numbers was via ratios: one could say that the ratio between this line and that line is as the ratio between, say, 1 and 2. But this does not mean that lines have numerical lengths (indeed, the ratio of 1 to 2 can be expressed as the ratio of 2 to 4, or 3 to 6, and so on). The realm of geometrical magnitudes and the realm of numbers were distinct.

The profound observations of Unguru and Rowe (1981, 1982) make it perfectly clear (in contrast to the then prevalent hypothesis) that classical Greek geometry was not a covert form of arithmetic or algebra, and that arithmetic and proto-algebra (for example, Diophantine algebra) were carefully distinguished from geometry by the Greeks. However, in a slow and gradual process, Greek geometry was historically arithmetized and algebraized, until the eventual emergence of an integrated algebraic geometry in the seventeenth century (Netz 2007; Corry 2013). The preceding phylogenetic story fits Dehaene's ontogenetic picture: we start with distinct circuits for number and space, but these circuits can be interrelated as we evolve.

However, this is not the only history out there. If we try to retrace the origins of Arabic algebra, and peel away the Euclidean justifications imposed on it by al-Khwarizmi and his successors, we find that Arabic algebra has two sources. One is the Babylonian quadratic calculations with rectangles, which derived the sides of rectangles from combinations of their sides, diagonal, and area. The other is the manipulations of unknown numbers in linear equations (possibly of Persian or Indian origin; see Høyrup 1990, 1998; Oaks and Alkhateeb 2005). So when we try to trace Arabic algebra back, we find that even after we separate it from the influence of Greek geometry, we have geometry integrated into its Babylonian source.

So the candidate for a pure arithmetic/algebra is the tradition of linear manipulations of unknown quantities (for example, given the price of 2 melons and 1 orange, and the price of 2 oranges and 1 melon, figure out the price of each). Scarcity of sources makes it difficult to say much about this tradition, but it is certain that linear manipulations of unknowns are not independent of geometrical measurements. Even if the motivation for these techniques was purely commercial, such obvious applications as pricing fabrics or land plots conflate the unknown arithmetical quantity with the way one quantifies fabric or land: according to their geometric length or area (this narrative is further complicated by bringing up another question: given the narrative of chapter 2 , should we think of money as a distinct quantitative category, or simply as number?).

So at the root of Arabic algebra, we find a conflation of arithmetic and geometric magnitudes, whereas in Greece we find a careful distinction between geometry and algebra. How can we put these stories together? The missing link is the practical mathematics that thrived in Greece as elsewhere, evolving from crafts, architecture, engineering, land measurement, and astronomy, where lines and numbers had always been conflated (Asper 2008; Netz 2007, 113). If this was the case, then the supposedly pure geometry and pure arithmetic of the Greeks were intellectual abstractions that grew from a world of mixed practical measurements of time, space, and cardinality. Now this phylogenetic narrative sounds much more like Walsh's ontogenetic story.

History didn't leave enough evidence to pinpoint the very moment of discovery or invention of concepts of magnitude. But again, as in the ontogenetic context, we must accommodate the possibility that the question is not well posed, and that no such moment existed. Indeed, the very notions of spatial magnitude and number may have emerged from a history of mixed and separated representations of various practices involving correlated and uncorrelated manifestations of space, time and cardinality. Notions of magnitude and number may have become *notions* "too late" in history—too late, in the sense that they become stable and unified cultural artifacts only after a long history of partly mixed and partly segregated, partly abstract and partly concrete forms of representations for specific practical contexts.

More precisely, if a unified concept of number emerged from coordinating different kinds of counting practices (some languages have different counting systems for animate objects, inanimate objects, money, time, ordinals—even English still fails to pluralize some countable entities, such as fish and sheep, and uses different words for cardinal and ordinal numbers), if a unified concept of spatial magnitude emerged from coordinating different kinds of measuring practices (using different terms, different bases, different upper limits, and different levels of precision), and if some of these counting practices had been mixed with measuring practices, while others were strictly segregated, then there's no unequivocal answer to whether numbers and spatial or temporal magnitudes, when each finally became a distinct linguistic-cultural notion, should be thought of as distinct or correlated.

Mathematical Metaphors

In the last few paragraphs, I have already committed a methodological *faux pas*. I projected on a history of evolved mathematical practices a neuro-cognitive debate that relates to the most elementary processing of magnitudes. I allowed myself this transgression because both debates seem to be organized around the question of modularity: is arithmetical processing an abstract and independent module, or is it something that emerges from nonspecialized quantitative knowledge and anchored to specific representations? I tried to explain why the question might not be well posed. I argued that the emergence of an arithmetical domain may arrive "too late," both onto- and phylogenetically, if by the time the domain stabilizes cognitively or culturally, it already conflates some specialized and some nonspecialized subpractices or brain circuits, which, considered in isolation, are not enough to count as self contained mathematical domains.

In order to relate the history of mathematics to neuro-cognition in a more viable manner, we must look into theories that explicitly engage the formation of higher mathematical concepts. But, as we will see, the context of higher mathematical cognition will raise concerns similar to those that emerged in the context of the number sense. The previous section, then, prepares us for the debate that follows.

The most widely acknowledged cognitive theory of higher mathematics is the theory of cognitive metaphor as applied to mathematical concept formation by Lakoff and Núñez (2000). This theory depends on the notion of "structured connectionism" in the brain, according to which "[n]euronal groups (of size, say, between, 10 and 100 neurons) are modeled as "nodes" which are meaningful and which enter into neural computation" (Lakoff 2008, 18).

Nodes have to do with mental simulations of embodied actions: "A meaningful node is a node that when activated results in the activation of a whole neural simulation and when inhibited inhibits that simulation" (Lakoff 2008, 19). These nodes give rise to "mental spaces." Each mental space is "a mental simulation characterizing an understanding of a situation, real or imagined. The entire space is governed by a gestalt node, which makes the mental space an 'entity' which, when activated, activates all the elements of the mental space" (Lakoff 2008, 30). Lakoff's examples of mental spaces include the understanding of such situations as "the U.S. during Clinton's presidency" and "a day in the journey of a fabled monk."

The notions of simulation and mental space are dominated by an objectification of action-processes: an action is made to correspond to a localized entity, dominated by a unified object—the gestalt node. Note, however, that a single neuron may partake in different nodes or spaces. Therefore, the theory does allow for a certain relaxation of the hierarchical structure of objectified action processes.

Moreover, what the theory of cognitive metaphors discusses is not so much "mental spaces" as "conceptual domains." A concept is defined as "something meaningful in human cognition that is ultimately grounded in experience and created via neural mechanisms" (Lakoff and Núñez 2000, 48), so it is grounded in mental spaces. Examples for conceptual domains include "physical putting in," "love," "travel," "the butcher stereotype," and so on. In the context of mathematics, domains include "concrete object collections," "numbers," "geometry," "sets," and so on. But a definition that characterizes the scope of intellectual domains and provides criteria for individuating them (saying where one ends and another begins) is lacking. The literature commonly defines a conceptual domain as "any coherent organization of experience" (Kövecses 2002, 4). This is obviously a very vague definition, which is hard to use for characterization or individuation. I will return to this vagueness later.

The basic notion of the theory of cognitive metaphors is conceptual metaphor:

> a *grounded, inference-preserving cross-domain mapping*—a neural mechanism that allows us to use the inferential structure of one conceptual domain (say, geometry) to reason about another (say, arithmetic). (Lakoff and Núñez 2000, 6)

Lakoff explains that "[i]nferences occur when the activation of one meaningful node or more results in the activation of another meaningful node" (Lakoff 2008, 19). Metaphors are instantiated in the brain by metaphor circuits, which activate links that make target domain nodes fire when source domain nodes fire. Since a certain combination of source domain activations can cause a source domain inference activation, and since the metaphor circuit further activates corresponding nodes in the target domain, an inference can be transferred from the source domain to the target domain. So, if we have a "number is a point on a line" metaphor, the inference "any two points have another point between them" can yield the inference "any two numbers have another number between them."

In fact, while this is not stated explicitly in the definitions quoted earlier, Lakoff and Núñez's conceptual metaphors can carry not only inferences, but also mathematical entities from a source domain to a target domain. For example, the diagonal of a square with unit side can be carried into the realm of arithmetic via a "magnitude is number" metaphor and become $\sqrt{2}$, even if before this metaphorical transfer arithmetic had no corresponding entity.

Note that metaphors are not necessarily symmetric—the neural activation may be one directional. Moreover, even when thinking in the direction of the metaphor, a target domain inference that is inhibited by target domain knowledge (that is, inconsistent with the target domain) will not be forced by a metaphor from the source domain. So the metaphorical transfer of inferences may be partial. In fact, the entire metaphor may be inhibited in some situations. Indeed, according to the theory, a metaphor circuit is governed by a gestalt node whose inhibition can prevent the metaphor from applying. This architecture provides the theory with fail-safe mechanisms that allow it to be adapted to empirical exceptions.

The next major notion of the theory is conceptual blend:

> the conceptual combination of two distinct cognitive structures with fixed correspondences between them.... This blend has entailments that follow from these correspondences, together with the inferential structure of *both domains.* (Lakoff and Núñez 2000, 48)

In neural terms, "A blend is an instance of one or more neural bindings" (Lakoff 2008, 30), where a binding "is responsible for two or more different conceptual or perceptual entities being considered a single entity.... It is not known just how neural binding operates in the brain. One hypothesis is that neural binding is the synchronous firing of nodes" (Lakoff 2008, 20). The difference between metaphor and binding is the difference between explaining integer addition in terms of stepping along the number line (using the "number is a point on a line" metaphor) and the Cartesian plane, where points and pairs of coordinates are considered to be a single entity, mentally bound together. Mental binding, like metaphor can also be conditioned or relaxed.

Examples of a conceptual metaphor and a conceptual blend are available in the following analysis of the number line (table 5.1). According to Lakoff and Núñez, the number line is a combination of (1) a conceptual blend of the continuous line and a set of elements resulting in a conception of the line as the set of its points, and (2) a conceptual metaphor projecting this blend on arithmetic, importing geometrical and set theoretical inferences to the realm of numbers.

My attraction to mathematical metaphors should be obvious. I have explored throughout how knowledge is transferred among domains to form mathematical practices (economy and numbers, analytic functions and formal power series, gender stereotypes and graph theory, and so on). But given that metaphors and blends can be conditioned and relaxed, it may seem that the entire theory is trivial. You can suppose any metaphor you want. Whenever a metaphor fails to work, you could simply claim that it's "inhibited" and ignore it. But the theory does include some components that safeguard against such triviality.

Table 5.1. Numbers Are Points on a Line (Fully Discretized Version)

Source Domain		Target Domain
The Space-Set Blend		
Naturally Continuous Space: The Line	Sets	Numbers
The line.	A set.	A set of numbers.
Point-locations.	Elements of the set.	Numbers.
Points are locations on the line.	Elements are members of the set.	Individual numbers are members of the set of numbers.
Point-locations are inherent to the line they are located on.	Members exist independently of the sets they are in.	Numbers exist independently of the sets they are in.
Two point-locations are distinct if they are different locations.	Two set members are distinct if they are different entities.	Two numbers are distinct if there is a nonzero difference between them.
Properties of the line.	Relations among members of the set.	Relations among numbers.
A point O.	An element 0.	Zero.
A point I to the right of O.	An element 1.	One.
Point P is to the right of point Q.	A relation $P > Q$.	Number P is greater than number Q.
Points to the left of O.	The subset of elements x, with $0 > x$.	Negative numbers.
The distance between O and P.	A function d that maps (O,P) onto an element x, with $x > 0$.	The absolute value of number P.

Source: Adapted from Lakoff and Núñez (2000), 281.

First, it is crucial for the theory that conceptual metaphors emerge from ground metaphors. Ground metaphors carry into mathematical domains (for example, arithmetic) such embodied experiences as forming collections of objects, subitizing (perceiving small quantities without counting), and matching collections in a one-to-one correspondence. Anchored in ground metaphors, the chain of mathematical metaphors is uniformly directed: from the concrete to the abstract, and not vice versa.

Another criterion that makes the theory of metaphor less trivial is the criterion of best fit.

What determines "fit"? Maximizing the number of overall neural bindings, including context and overall knowledge, without contradiction, that is, without encountering any mutual inhibition. A node A fits a complex network B better than complex network B′ if the strength of neural bindings one can create between A and B without mutual inhibition is greater than with B′. (Lakoff 2008, 24)

The principle of best fit is derived from Hebb's law, which in Carla Schatz's catch phrase reads: "Cells that fire together, wire together." Bad metaphors, which require frequent inhibitions (namely, many cases where corresponding nodes are inhibited from "firing together"), won't produce strong neural links. This yields a sort of natural selection among metaphors.

The test for the theory of mathematical metaphor is therefore its ability to account for as much mathematical knowledge as possible, while preserving as many inferences as possible, and building from the concretely embodied to the abstract.

Some Challenges to the Theory of Mathematical Metaphors

The reception of the theory of mathematical metaphors was rather critical. Reviewers claimed that it was poorly anchored in empirical cognitive science and anthropological observations (Dubinsky 1999; Goldin 2001; Madden 2001; Ernest 2010); that its philosophical argument was an attack against straw-men (Gold 2001; Henderson 2002); that it failed to distinguish different layers of mathematical knowledge (Sinclair and Schiralli 2003; Ernest 2010); that its attempt to draw unique metaphorical lineages of mathematical ideas was too narrow, reflecting post hoc textbook reconstructions rather than historical cognitive processes (Dubinsky 1999; Goldin 2001; Sinclair and Schiralli 2003; Schlimm 2013); and that conceptual metaphors can only account for a limited portion of mathematical knowledge formation, as they neglect cognitive processes such as generalization, abstraction, formal manipulation, invariance, inversion, and metonymy (Dubinsky 1999; Sinclair and Schiralli 2003; Ernest 2010; Mason 2010).

Despite these critical observations, the book is often cited and highly influential. One of the reasons for its success is its fundamental insight: it makes sense to think about mathematics in terms of a transfer of ideas between domains. But the book's concept of metaphor is far too thin and rigid to account for mathematical knowledge. In other words, the constraints that it imposes on mathematical practice are far too strong to reflect any really existing mathematical culture.

In this section, I will challenge the theory on several grounds. I will first argue that the notion of "best fit" is severely underdetermined, and cannot impose the clear and distinct hierarchical organization that the theory wishes to impart. Then I will challenge the notion of "conceptual domain," and argue that the architecture of conceptual domains must be thoroughly revised. Finally, I will challenge the directionality of the theory (from the concrete to the abstract) and argue that the entire theory depends on metaphorically imposing an abstract formal mathematical model on concrete mathematical practice. The next chapter will provide some historical case studies to further establish these critiques.

Best Fit for Whom?

The following vignette is brought by Lakoff and Núñez to show how some metaphors are more "normal" than others—that is, provide a better fit with the various connections between embodied experience and other mathematical metaphors. Specifically, they explain an apparent paradox by suggesting that the sense of paradox is imposed by a mathematical metaphor that is in conflict with everyday experience. They point out that a different choice of metaphor would dissolve the sense of paradox.

The apparent paradox—a well-known folk-mathematical curiosity, appearing in a different version in a letter written in 1931 by Luzin to Vygodskii (along with a beautiful account of the reification of infinitesimals from the perspective of a struggling student; see Demidov and Shenitzer 2000)—concerns the following situation. Consider a sequence of bumpy curves drawn over a horizontal unit segment. Each curve consists of a horizontal row of connected congruent half circles. The radii of the half circles diminish as we move from one curve to the next. This sequence of curves converges uniformly to the unit segment (see figure 5.1), where "uniform convergence" means that the maximal distance between the horizontal segment and the points above it in a given curve tends to zero as the sequence of curves unfolds. But despite the uniform convergence of the curves to the segment, the length

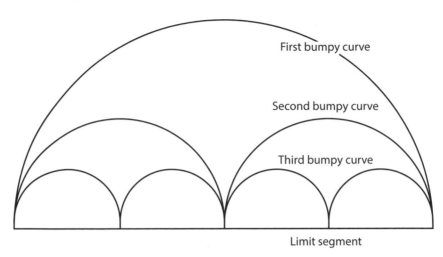

Figure 5.1: The bumpy curves, no matter how "flat" they become, are always of length π/2, whereas the straight limit segment (0,1) is of length 1. Revised from Lakoff and Núñez (2000), 329.

of all the curves remains π/2, whereas the length of the limit segment is 1. In other words, despite the convergence of the curves to the segment, the lengths of the curves do not converge to the length of the segment.

Lakoff and Núñez explain this apparent paradox by claiming that the metaphors involved in conceiving the uniform convergence of curves violate our expectations. Our "normal expectations" assume that length is an inherent property of a curve, and entail that "[n]early identical curves should have nearly identical properties," and so nearly identical lengths (Lakoff and Núñez 2000, 330). Therefore, according to the authors, uniform convergence is not a "normal" conceptualization of near identity of curves. In other words, it violates the requirement of best fit.

But why consider length an "inherent property" of a line, as opposed to other, supposedly accidental, properties? According to Lakoff and Núñez, this follows from our experience of measuring curves (such as the bumpy curves here) ever more precisely by using ever shorter measuring sticks. Given this measurement practice, we only feel that the bumpy curves are nearly identical to the limit (here, straight) line segment when the ever-shrinking sticks we use to measure them (that is, the tangent segments approximating the bumpy curves) become nearly parallel to the sticks we would use to measure the limit (straight, horizontal) line segment. But in figure 5.1, this criterion is violated: the measuring sticks for the bumpy curves would keep changing their directions as we move along the curves (even when the bumps are very small), and would never remain nearly parallel to the horizontal limit line. Therefore, saying that the bumpy curves converge to the straight line is—albeit mathematically tenable—a confusing metaphorical use of the concept of near identity or convergence; hence the sense of paradox.

To respect our normal expectations and avoid the confusion, our convergence concept for curves should be, according to Lakoff and Núñez, the so-called C^1 convergence, where both the points on the curves *and* their tangents (or derivatives or directions of short measuring sticks) are nearly those of the limit segment. If we were to use *this* metaphorical extension of the concept of convergence, we would not say that the bumpy curves converge to the straight line, and no sense of paradox would ensue.

But I, for one, have never measured a curve with shrinking measuring sticks. I doubt that this "embodied experience" is very often actually embodied. For example, if the bumpy curve represented a fence, then uniform convergence would definitely fit the experience of near identity between the fence and ground from the point of view of someone who tried to jump over this fence. Seamstresses may have a different experience of nearly identical curves, say when they sew a thread in loops around a hemline. In that case, the length of the approximating curve (the thread) is "normally" expected to exceed the length of the approximated curve (the hemline) even if the former is sewn very tight, almost superposed over the hemline—and no sense of paradox ensues. Another "normal" expectation may be that of someone pushing a cart along a road shaped like the bumpy curves. There, the second derivative has to do with the effort required to push the cart, and is likely to be a crucial component of embodied experience, precluding an experience of near identity if we had only C^1 convergence. The distinction between normal and abnormal uses of metaphors turns out to be rather contingent.

This example shows that metaphorical "best fit" depends on context, and may change with the particular embodied practices of different people. However, if, like Lakoff and Núñez, we claim that the "precise characterizations given of metaphorical mappings, blends, and special cases reveal real, stable, and precise conceptual structure" (Lakoff and Núñez 2000, 375), then one indeed has to choose a stable and precise "normal" core, and brand anything that deviates from it as abnormal.

To obtain such a "normal" core, one indeed would have to declare that "Behavior, performance, and competence of particular individuals are secondary" (Lakoff and Núñez 2000, 111). It is only then that one could exclude form the discussion of "mathematics itself" the occasional oddball (for example, Cantor, who nevertheless was analyzed by Núñez 2005) or struggling schoolchildren (whose metaphors are insightfully analyzed by Presmeg 1992). But then we rob mathematical metaphors of one of their most productive features: their open-ended interpretability (Dubinsky 1999) and their useful ambiguities (to borrow Emily Grosholz's term) as explored in this book.

This "normalization" is not a marginal aspect of the theory of mathematical metaphors. In the next chapter, we will see a similar attempt to normalize mathematical thought in the context of mathematical infinities, which Lakoff and Núñez strive to reduce to a single metaphor. (Schlimm [2013] offers a similar critique in the context of set theory.) And while Lakoff and Núñez reconstruct a single trajectory for the formation of the concept of number, Ernest insists that "there is not a single, uniquely defined semiotic system of number, but rather a family of overlapping, intertransforming representations constituting the semiotic systems of number" (2006, 94). The theory of mathematical metaphor, in its current form, risks impoverishing mathematics and imposing on it unnecessary normative constraints.

What Is a Conceptual Domain?

In my historical analysis of geometry, arithmetic, and algebra, I suggested that a narrative starting from the primitive domains of geometry and number, later blended into algebraic geometry, is problematic. Geometric and arithmetic treatments of magnitudes have always been subject to a complex negotiation between integration and segregation, and the same is valid for other case studies considered in this book. Instead of individuation criteria, conceptual domains have varying levels of separation or integration in given historical and cultural settings, gleaned pragmatically from the codependence of their respective practices.

This account of conceptual domains precludes a clear and distinct hierarchy evolving unidirectionally through metaphors and blends. But if such a structure is precluded, what is it between which conceptual metaphors can take place?

Mowat and Davis (2010) suggest that instead of hierarchically organized conceptual domains, we could think of mathematics in terms of a nonlinear network of concepts connected by superimposing metaphors. They further suggest that the more metaphorical links one establishes between concepts, the more robust the mathematical network becomes. Instead of a competition between different metaphors over "best fit," the robustness of the structure would follow from the interaction of many different, even partly contradictory, metaphors.

This image has an interesting impact on the relation between conceptual domains, as the exclusion of unidirectional hierarchy allows for circularities. For example, a domain A, metaphorically linked to a domain B, which is in turn metaphorically linked to C, which links again to A, generates a circular relation between domains (which may explain the hesitating integration/segregation of, say, arithmetic and geometry). In fact, C may even be a subdomain of A, violating even internal hierarchies, as the entire domain may be metaphorically shaped by its subdomain and vice versa.

But this still does not sort out the problem of individuating domains. We may indeed find that a conceptual domain is itself composed of a network of linked subdomains similar in structure to the super-domain, and these subdomains, in turn, may turn out to consist of similarly structured and linked subdomains, and so on. But can we truly imagine mathematical concepts as endlessly decomposable with no final constitutive elements? This seems to make no sense. If mathematics takes place in an embodied neural mechanism, then a mathematical domain must correspond to a bunch of neurons, and as we descend into subdomains, we must hit some minimal functional neural structure. In any sequence of embedded Russian *Matryoshka* dolls, there must always be a smallest doll.

But this holds only if we think inside the box. If our thinking is not something that happens strictly inside neural components enclosed in the boxes mounted over our shoulders, but something that depends on the interaction between neurons and a world, then things can turn out differently.

Several theses relate to the title "embodied cognition" under which the theory of mathematical metaphors positions itself. One of them is "offloading cognitive work onto the environment" (Wilson 2002; Pfeifer and Bongard 2006). We do not compute strictly inside our heads, but use the environment for calculation. Think, for example, of trying to move a sofa through a staircase: we set it in an angle that seems to be more or less OK, push until we bump into something, readjust, backtrack—that is, offload computational maneuvers onto the environment instead of calculating a trajectory in advance. In a more strictly mathematical context, Landy et al. (2014) show how cognitive work related to elementary algebra (for example, which operation in a formula to perform first) is offloaded onto the setting and spacing mathematical signs.

Now, even if our brains do have innate minimal mathematical neural structures (such as Dehaene's or Walsh's circuits), changes in our signs, tools, and environments refute the assumption of stable minimal embodied quantity components. Indeed, a minimal mathematical component would be the application of a minimal neural structure to an element of the environment. But our environment does not reduce to minimal components determined once and for all—humans reform their signs, tools, and worlds as they go along.

De Cruz and de Smedt (2010) point out that the transition from imprecise innate mathematical abilities to more precise formal ones is not a rigid isomorphism similar to Lakoff and Núñez's metaphor tables. This transition has to do with turning imprecise innate abilities into a more rigid arithmetic system via signs, tools and practices of measurement and inscription involving body parts, tokens, words, or written symbols (De Cruz 2008; see also Mason's (2010) note on the role of cards in embodying the notion of permutations and Ursula Klein's (2003) notion of "paper tools"). They even suggest evidence that the means we use to represent numbers affect the processes inside our brains and reform their function.

In chapter 2, we saw how mathematics was changed by economic practices, and in chapter 4, we saw how gender signs helped form mathematical knowledge. The historical treatment of algebra and geometry in the next chapter further demonstrates how mathematical metaphors depend on signs and tools. I mention here only briefly that Descartes's mathematics is intimately related to his new tools for constructing curves, without which he could not have solved some previously intractable algebraic problems (Bos 2001). The same is true for Bombelli's right-angled-ruler-based constructions and, more generally, his engineer's perspective (Wagner 2010b). Piero della Francesca, who wrote notable mathematical treatises, is much better known as a Renaissance artist, and his embodied mind was immersed in quantitative spatial representation techniques (Field 2005). Arithmetic, algebra, and geometry are not only a cognitive and conceptual blend; they're a blend that depends on the tools of the trade and the way they shape minds.

Indeed, even the most basic, supposedly universal tools change their role depending on what is counted, for what purpose, and in which context. Subitizing and finger counting in song-play is different from if "your life depended on it," to use Walkerdine's (1977) expression for children street peddlers. The semiotic role of fingers and their use in calculation also differs in various elementary everyday contexts (Walkerdine 1977, 67).

Lakoff and Núñez recognize a historico-technological dimension of mathematics (2000, 359–62), but this dimension hardly ever comes up in their discussion. Indeed, they anchor mathematics to an embodied mind that has hardly anything to do with the vicissitudes of practice. Mathematics involves our bodies, but since these bodies cannot *do* precise mathematics without signs and tools, and since the choice of signs and tools is open ended, the limits of conceptual domains are more fluid than Lakoff and Núñez allow for.

A conceptual domain is, therefore, not a minimal unit that's located in the brain or even in the naked body. It is a fluid combination of neural components, bodies, signs, and tools. Indeed, if neural mechanisms, signs, and tools are kept apart, no mathematical work can be done.

Mathematical metaphors operate not in a rigid hierarchical system of domains, but in a network of fluid, open ended temporary constellations of neurons, bodies, signs, and tools.

In Which Direction Does the Theory Go?

The theory of mathematical metaphors assumes that they are grounded in the body. This means that they necessarily go from the concrete to the abstract. In particular, since geometry represents concrete spatial experience and numbers represent collections of objects, while algebra is a symbolic abstraction, metaphors should go from the former to the latter, not in the opposite direction.

In reality, this is not always the case. The discussion of geometry and algebra in the following chapter will prove that inferences sometimes go from algebra to geometry. But here let us consider the following example. One famous eighteenth-century identity, endorsed by, among others, Euler and Leibniz, is the following:

$$1 - 1 + 1 - 1 + \ldots = \tfrac{1}{2}.$$

One way to derive it is to substitute 1 for x in the series

$$1 - x + x^2 - x^3 + \ldots = 1/(1+x)$$

(Jahnke 2003, 121–22; Sandifer 2007, ch. 31). This derivation has nothing to do with the semantic notions of the numbers involved—it was a symbolic derivation, carrying inferences from algebra to arithmetic. This kind of reasoning was used for solving geometrical problems dealing with lengths and areas, so it was transferred to geometry as well, violating the supposed directionality of conceptual metaphors.

This last example is important precisely because of its questionable status. There are obviously many similar symbolic derivations that are accepted in contemporary terms as well, but this example highlights the instability of mathematical truths. Today, the preceding identity will be mostly rejected, unless one brings up the notion of Cesàro summation (the limit as n goes to infinity of the average of the first n partial sums). In that case, the identity is justified because we provide it with a semantic content. Does this suggest that we only accept a formal inference if it can be grounded from "below" as well?

Well, this is not just how it works. If we look at Euler's justification, we see that he accepted such maneuvers only if they fit strongly into a network of algebraic practices. For example, if the same result could be derived by various different means (even if all those means are invalid by today's standards), then Euler would be more likely to accept it. Sums of infinite series did not have to relate to quantitative approximations, but to applications of analytic algorithms to analytic expressions (Ferraro 2007). In the words of Lagrange: "This value [of a function] will be truly represented by [its Taylor] series, but the convergence of this series will depend on i [the value of the variable, corresponding to x earlier]" (Lagrange 1797, 67). Limit value and analytic value were not considered to be the same thing, but, according to Daniel Bernoulli, "one cannot challenge the exactitude of such a substitution without overturning the most common principles of analysis" (quoted in Bottazzini 1986, 53).

These techniques did lead to inconsistencies, but these monsters could be contained or warded off. It was only later, in the nineteenth century, as new techniques entered the field, that inconsistencies became too widespread, and attitudes changed. Transfers of inferences (or metaphors) that enjoyed a good fit in the eighteenth century, were no longer manageable in the nineteenth. The reaction was to prefer the identification of the sum of a series with the limit of its partial sums. However, as we saw in chapter 4, formal approaches to power series survive to this very day. This is a fine example of negotiating the constraints of new standards of rigor and preservation of older knowledge.

New ideas and structures often occur in mathematics based on formal analogies rather than grounded metaphors. The subsequent endorsement or rejection of formal analogies depends on their integration with other mathematical practices. They must present a good fit with existing knowledge, while not violating the constraints on rigor that maintain mathematical consensus. When these constraints conflict, new reconstructions, superpositions, and deferrals of interpretations ensue in order to reduce the tension between competing constraints. But this process does not have to be metaphorically grounded in more concrete mathematical concepts. It has to work well with the structure as a whole.

Mathematicians do seek intuitions when engaging with new mathematics, and these intuitions often have a concrete character, but this is not a necessary requisite. For example, the representation of complex numbers as points in the plane helped promote the endorsement of complex numbers that evolved from purely formal manipulations of roots of negative numbers, first considered as "sophistic" entities (see chapter 2). But the integration of complex numbers into mathematics did not *depend* on their geometric representation; it depended on their successful cohabitation with general analysis and algebra.

In fact, formal analogies are not only part of what keeps mathematics moving; it is also part of the engine steering the theory of mathematical metaphors itself. Indeed, while trying to semantically ground mathematical concepts back in embodied experience, Lakoff and Núñez depend on syntactic analogies. For instance, the metaphor "a number is a collection of objects" is supposed to carry multiplicative associativity from concrete object collections to abstract arithmetic.

Source inference:

> Pooling A collections of size B and pooling that number of collections of size C gives a collection of the same resulting size as pooling the number of A collections of the size of the collection formed by pooling B collections of size C.

Target inference:

> Multiplying A times B and multiplying the result times C gives the same number as multiplying A times the result of multiplying B times C. (Lakoff and Núñez 2000, 63)

But is the source inference really more concrete? Is it even intelligible? Since the source inference requires the rather unintuitive practice of turning a complicated numbered structure (A collections of size B) into tokens that number other collections, it is rather difficult to decipher the experience underlying the "source" inference, and why it should feel true. In fact, the easiest way to understand the "source" inference is to project the arithmetic target inference onto collections of objects.

Similarly, for the unlimited iterability of addition, the origin inference is: "You can add *object collections* indefinitely," and the target inference is "You can add *numbers* indefinitely" (Lakoff and Núñez 2000,

58). But objects of embodied experience are precisely those that cannot be added indefinitely (you can't always have another apple—they may have run out!). Our sense of an indefinite addition arises, if at all, from formal arithmetic experience. It is not grounded in experience, but projected on experience.

This most elementary metaphor (arithmetic as dealing with object collections), which works great in elementary schools, works precisely because it is loose (see Presmeg 1992). It becomes rigorous only if we project onto our experience with object collections our formal experience of numbers, and the supposed grounding is allowed to become circular. Rather than treating embodied experience as an underlying generative ground, the theory of mathematical metaphor imposes a post hoc mathematized reconstruction of embodied experience.

In fact, the entire theory of mathematical metaphors is a formal structure that imposes its neat hierarchies on mathematical practice. Indeed, to explain the philosophical formalist approach to mathematics on their terms, Lakoff and Núñez reconstruct the following metaphor:

On the right-hand side of table 5.2 are mathematical ideas—that is, what the theory of conceptual metaphor deals with. On the left-hand side are formal interpretations that Hilbertian formalists (presented in the first section of chapter 1) project on mathematical ideas—a reduction of ideas to combinations of empty signs, which Lakoff and Núñez consider inadequate. This table is meant to explain how formalism came to be.

But I believe that this table actually serves to explain the problematic origins of the theory of mathematical metaphors. Indeed, something in the neatness of the correspondence is suspect. If mathematical ideas are fundamentally different from their formal counterparts (as Lakoff and Núñez claim), then how come the metaphor that translates them into systems of empty signs fits so well? Indeed, this metaphor fits so well, that some critics failed to distinguish mathematical metaphors from formal isomorphisms (Lakoff and Núñez 2001).

Here is one possible explanation for why the preceding metaphor works so well: Lakoff and Núñez's notions of "mathematical ideas" and "mathematical metaphor" might themselves depend on an unacknowledged metaphor—their own metaphor of metaphor—a metaphor reconstructing cognition and its metaphors in the image of none other than the structures and isomorphisms of mathematical formalism! They

Table 5.2. The Formal Reduction Metaphor

Source Domain	Target Domain
Sets and Symbols	Mathematical Ideals
A set-theoretical entity (for example, a set, a member, a set-theoretical structure)	A mathematical concept
An ordered n-tuple	An n-place relation among mathematical concepts
A set of ordered pairs (suitably constrained)	A function or an operator
Constraints on a set-theoretical structure	Conceptual axioms: ideas characterizing the essence of the subject matter
Inherently meaningless symbol strings combined under certain rules	The symbolization of ideas in the mathematical subject matter
Inherently meaningless symbol strings called "axioms"	The symbolization of the conceptual axioms—the ideas characterizing the essence of the subject matter
A mapping (called an "interpretation") from the inherently meaningless symbol strings to the set-theoretical structure	The symbolization relation between symbols and the ideas they symbolize

Source: Adapted from Lakoff and Núñez (2000), 371.

take mathematical practice, and project on it entities and inferences that come from the realm of formal structures ...

Lakoff and Núñez's theory manages to account only for a formal/structural version of mathematics, a mathematics reduced to a hierarchical system of structures emerging from axioms (grounding inferences) and partial isomorphisms (metaphors). This reduced version is not identical to that of mathematical formalism, but is homologous to it (it's no surprise, then, that cognitive metaphors can be so successfully represented as a hierarchical computerized database; see Lakoff 2008, 36–37). But both the formalist and the cognitivist reductions try to impose untenable constraints on mathematics, and are not adequate descriptions of mathematical practice.

So How Should We Think about Mathematical Metaphors?

It's time to put things together, and suggest an alternative to Lakoff and Núñez's definition of mathematical metaphor. The original definition was: "a *grounded, inference-preserving cross-domain mapping—a* neural mechanism that allows us to use the inferential structure of one conceptual domain (say, geometry) to reason about another (say, arithmetic)" (Lakoff and Núñez 2000, 6).

So let's review the definition term by term, and attempt to revise it. The first term, "grounded," refers to embodied experiences with a strong tendency to prefer the innate and universal. I suggest a focus on embodied experiences more inclined toward contingent practices with signs and tools.

Second, metaphors are qualified as "inference preserving." But as we'll see in the next chapter, inferences are not all that metaphors transmit. They transmit entities, inferences, formations of problems, means of representation and solution, epistemological status—and this list is not meant to be exhaustive. I think that the structuralist term "articulation" (Barthes 1971) can more or less cover this field. "Articulation" refers to the relative partition of a phenomenon into elements and to the determination of relations between those elements, such as difference and repetition or conjunction and exclusion. Articulation covers the individuation of entities, their primary logical relations (for example, inference), their secondary organization into operational units (for example, problems), their relations to external structures (for example, representations), and their place within meta structures (for example, epistemological relations). So instead of inference preservation, we should talk about relative articulation.

Third, according to my preceding analysis, domains are not only neural units, but interactions of neurons, bodies, signs, and tools, which must remain open on several dimensions: open with respect to each other (a neural structure interacting with different systems of signs; a sign involved in different practices); open with respect to the outside (what goes on outside the classroom or mathematical text does interfere with what goes on inside); and open with respect to themselves (signs, practices, and neural structures always undergo some slight variations in repetition, and it is never absolutely settled when these variations escape a given framework). Following Derrida (1988), I will use the term "context" for such underdetermined and ambiguously bounded domains.

So our definition of metaphor becomes: "relative articulation across mathematical contexts—an embodied practice with signs and tools that draws on one context to articulate another." This is not empty jargon. It can be explicated in tables like Lakoff and Núñez's table 5.1 earlier that sought to represent the ideas behind Cartesian geometry. But these tables will be messy, complicated, and not very well tabulated. If we were to properly account for early modern geometric algebra, for instance, these tables would present the complex rearticulation of Arabic algebra and Greek geometry in the contexts of Renaissance economic and juridical discourses (among others), forming Italian and French geometric algebra (see chapter 2 and the first section of chapter 4; Wagner 2010b, 2010c; Cifoletti 1192, 1995). If we settle for Lakoff and Núñez's narrower definition, mathematical metaphor would cover only a formal skeleton of algebraized geometry.

Finally, I should emphasize two reservations. First, I don't consider mathematical metaphors to be the only mechanism through which mathematics is formed. I think that metaphors are an important part of how mathematics evolves, but do not monopolize mathematical practice. Second, even in the restricted context of exploring mathematical metaphors, I believe that the proper task of humanist research of mathematics is not simply to construct intricate tables of how different mathematical contexts are relatively articulated. I believe that our task is to constantly problematize notions of metaphor so as to reflect on how and why, in specific historic, social and practical circumstances, some transfers of knowledge survive the break with their contexts, while others end up rejected, marked as invalid and false. This should be done by relating metaphors to the wider systems of natural, cognitive, social, and other constraints that apply to mathematical practice.

An Alternative Neural Picture

While my arguments may convince some readers that mathematical cognition is much more complex than the rigid picture assumed by the theory of mathematical metaphors, it is extremely vague from a neurological point of view. Fortunately, some neuro-cognitivists offer models that fit better the complex picture that I try to paint. Here I will follow the work of Walter J. Freeman III.

Here's how Freeman presents the mainstream neuro-cognitive paradigm, which he rejects:

[W]hen sensory input excites receptor neurons, their pulses represent primitive elements of sensation, or features. The primary sensory cortex combines these representations of features into representations of objects and transmits them to adjacent association areas; for example, a combination of lines and colors might make up a face, a set of phonemes might form a sentence, and a sequence of joint angles and tissue pressures might represent a gesture. They believe that representations of objects are transmitted from the association cortices to the frontal lobes, where objects are abstracted into concepts to which meanings and value are attached. Somehow, in these chains of reaction, a sensation becomes a perception, but they haven't been able to show where that happens, or in what way a perception differs from a sensation, or where the information in a perception changes into the information in a command for a behavior. (Freeman 2000, 98)

According to this picture, some stimuli are sensed and coordinated into distinct objects (say, numbers) using some system that is in line with Dehaene's or Walsh's models. Then, these objects may be interrelated with other objects or experience modules to form more complex models in line with mathematical metaphor theory. Stimulated nodes further activate other nodes connected to them via links that represent inferences (in the sense of cognitive metaphor theory), and we derive conclusions.

In order to explore an alternative, we need to start with Freeman's research on smell in animals. When Freeman tried to figure out what happens in animal brains when they sense an odorant, he did find certain regularities. But these regularities did not take the form of a neuron or a bunch of neurons "lighting up." The regularities manifested in the pattern of activity in a rather large area of the olfactory bulb—not a small microcircuit.

The activity pattern was expressed by a coordinated activity between all neurons in the sampled area. EEG recordings revealed a common wave pattern across the entire area. But the different locations in this area, which all presented the same pattern in terms of wave frequencies, differed in the amplitude of the wave (the "strength" of the wave, rather than its shape). So if one mapped the amplitudes across the area, one obtained a sort of topographical map of amplitudes (or "amplitude modulation") that served as a footprint for the specific odorant.

Now, several points must be emphasized concerning this finding (Freeman 2000, ch. 4):

- Not just any odorant produced a stable map. In order to get a stable amplitude modulation, the animals had to be trained to recognize the odorants by assigning them some reinforcement (reward, punishment). In other words, the smells had to be made meaningful for the animal.
- The activation of the amplitude modulation was not automatic even when a smell was meaningful. For example, a hungry cat responded to the smell of fish with a clear amplitude modulation, while a recently fed cat did not.
- The maps were more or less the same between subsequent exposures of the same individual to the same odorant in similar conditions, but differed from the patterns of the same odorant in other individuals.
- Once an amplitude modulation was established, the animal was trained to recognize other smells. When the original smell was reintroduced, the amplitude modulation for the original smell was changed. The intermediary experiences of the animal changed its response to an older meaningful stimulus.

There are several important consequences to these observations. First, the brain response is not determined by the stimulus. The same stimulus yields different amplitude modulations, depending on the context and the animal's history. Second, it is not determined by the sensory detectors: the same odorant may be sensed by different detector cells and pass through different routes on its way to forming the same modulation; on the other hand, even if the same sensory cells were involved, context and history could change the eventual result. Third, new reinforced stimuli did not just produce new amplitude maps, but transformed the maps of old meaningful stimuli as well. Therefore, the amplitude modulation is not a "representation" or "encoding" of the odorant. Instead, it is a response that integrated the stimulus, its meaning for the animal, the context and the animal's history.

How can we make sense of these findings in neural terms? Freeman suggests that we look at the activity of a brain area as a chaotic dynamic system. It has a rest or background state, and several "attractors"— namely, relatively stable activity patterns (in our case, amplitude maps) that the brain dynamics tend to and that the structure of the neural network can sustain for a while. When a strong enough stimulus is introduced, it upsets the rest stage, and initiates a process which may tend toward one of the attractors or back to the rest state.

When a stimulus is repeatedly associated with a meaningful reinforcement (pleasure, aversion), the resulting brain activation reshapes the interaction between neurons ("neurons that fire together wire together"). This changes not just the reaction to a specific stimulus, but the underlying neural connections, the dynamics of the network, and the totality of its various attractors. As a new attractor forms, old attractors may change.

To explain this in more intuitive terms, think for example of the water in a bay. If there's hardly any wind, the water will be calm. If the winds are stronger, certain patterns of currents and waves may form. These current and wave patterns are not unique. They depend, for example, on the direction of the wind, the season and the currents outside the bay. Still, a local fisherman knows the most likely stable and enduring patterns ("attractors") of his bay. Now, a wild storm can change the ground topography in such a way that the bay will change its shape, give rise to a new pattern of currents and waves, and change some features of the old patterns of currents and waves. Changing the ground formation in one side of the bay may affect currents and waves throughout.

In the context of the brain, the dynamical system involved in odor sensation is not isolated, but connected to the rest of the brain. Therefore, the general state of the brain (sleep, hunger, preparing for a certain action) serves as an operator that changes the behavior of the network, predisposing it in favor or against certain attractors (in terms of our last metaphor, one can think of high and low tide as an operator affecting the currents and waves in the bay).

A key notion in this context is "preafference": when an action is planned, messages are sent not only to motor systems, but also to sensory systems, to prepare them for the expected results of the action. These messages make it more likely that stimuli will drive the entire system toward the most relevant attractors or activity patterns. According to Freeman,

> Preafference provides an order parameter that shapes the attractor landscapes, making it easier to capture expected or desired stimuli by enlarging or deepening the basins of their attractors.... The same limbic message is sent to all the sensory cortices, so that the choice of a goal orients the senses in the same context, whether it is to find food, safety, or the feeling of power and comprehension that occurs when dopamine receptors are activated. The organism has some idea, whether correct or mistaken, of what it is looking for. (Freeman 2000, 108–9)

We can see here a strong phenomenological tendency in Freeman's thinking. Indeed Freeman and his colleague Robert Kozma write: "In our view perception begins with intention and not with sensation," and follow with a citation of Merleau-Ponty (response to Cohen Kadosh and Walsh 2009, 337). Another philosophical ally that Freeman (2008) found is Thomas Aquinas. The Thomist articulation of universal/singular (or, more generally, intellect/matter) appears to Freeman to reflect the distinction between the amplitude modulation and the material stimulus, and the Thomist "tending" of the intellect is interpreted by Freeman as relating to preafference (but I'm not sure that Freeman's "Godless" reading would be very popular among Thomas experts).

Just as with the transition from Dehaene's experimental data to the theory of mathematical metaphor, the transition from the experimental data concerning sensory cortices to a wider understanding of cognition is still exploratory and conjectural. According to Freeman, the sensory cortices' amplitude maps are an "early stage in the construction of meaning ... by which an animal 'in-forms' itself as to what to do with or about an odorant, such as whether to eat the food or run from the predator giving the odorant information" (Freeman 2000, 89–90). "Meaning" here is the integration of a stimulus into a goal oriented action. The sensory patterns are supposed to integrate with other brain patterns to generate ever wider brain dynamics that represent more complex meanings. What distinguishes humans in this picture is the capacity to construct hypothetical meta-operators that combine and reshape the ordinary wave packets that we share with other mammals and make symbols. These representations are in, on (e.g., tattoos), or outside the body, serving social planning and communication. (response to Cohen Kadosh and Walsh 2009, 337)

Let's explore what happens when we try to apply this theory to mathematics. First, Freeman rejects both Dehaene's and Walsh's attempts to locate number or magnitudes in small microsystems. According to Freeman, meaningful numbers can only emerge in the dynamics of larger brain areas, akin to his amplitude maps. He therefore rejects the focus in mainstream research on finding which neurons shoot and which stay silent

> I propose that every neuron and every patch participates in every experience and behavior, even if its contribution is to silence its pulse train or stay dark in a brain image. What is important is the small fraction of semi-autonomous activity in every part that is coordinated, not the small fraction of neurons or patches that is more active than the average. (Freeman 2000, 109–10)

If we're looking for cognitive categories,

> They are to be sought in large-scale neural structure and activity having low spatial density and wide correlation length. They are mesoscopic patterns that differ markedly from microscopic sensory and neural activity patterns, which are spatially localized at high density in small clusters of neurons. Microscopic activity driven by sensory input cannot be part of the knowledge base because it is unique and ephemeral; knowledge in the form of interrelated categories is enduring yet continually changing over the lifespan of the brain and body. (Freeman 2009, 3)

This analysis works well with several points that I made earlier. First, a concept is no longer an independent representation (possibly built upon an embodied experience, but then objectified and stabilized). It is an integration of embodied action with tools and signs, personal and cultural history and stimuli. An inference (the formation of a pattern in sensory circuits that triggers the formation of a pattern in action circuits) is no longer something that just happens because a node is lit, a metaphor is activated, and then another node is lit; instead, it is the result of a specific attunement predisposing us toward the inference.

Second, with the notion of preafference, the hierarchical grounding is no longer imposed. If an action can shape sensation, and not just the other way around, there's no problem carrying knowledge from more abstract activities (for example, algebra) to more concrete experiences (for example, embodied handling of object collections).

Third, the relation between mathematical concepts is inherently fluid. It is not because domain A is linked to domain B that they relate, but because they are both expressions of the dynamics of the same underlying neural network. If numbers and geometric magnitudes correspond to different modulation patterns in the same network, the carpet is pulled from under the domain individuation problem.

Fourth, a certain change in the overall system may change all concepts simultaneously, because a change in the structure of the neural network may affects all the attractors in the system at once. A certain reorganization of the underlying neural structure will affect, say, geometry and algebra at once, without waiting for an inference to occur in one domain and then move on to the other. Similarly, since different stimuli can direct the system toward the same attractor, the same meaning (expectations, behaviors) may emerge from different mathematical entities (numbers, points, and so on) without it first being applied to one, and then transferred to another. Of course, the latter process could still take place: a new arithmetical practice in one domain (say arithmetic) will reshape the entire neural network, and result in a related new geometrical practice.

Finally, an inference (triggering an action pattern by a sensory pattern) no longer has to follow a wiring that prepares for it in advance, as in the theory of mathematical metaphor. The metaphor theory assumes that the domains must already be bound or shoot together in order to be subsequently wired together by a metaphor that enables the inference. In Freeman's model, an inference (a new pattern that follows a certain stimulus) can also emerge accidentally (or, more scientifically, as a result of the chaotic dynamics of the system). The rewards or punishments associated with the inference (its adequacy with respect to the constraints that govern mathematical practice) will then help reinforce or inhibit the new inference or pattern in future instances. This aleatory component of reasoning is a better explanation of creative breakthroughs than the one offered by the theory of conceptual metaphor.

The observations in the last few paragraphs do not pretend to serve as a model that we can try to test empirically—this is the purview of scientists. But I believe that these observations offer a better allegory (or metaphor) for mathematical practice than the one derived from imposing on mathematical cognition the ordered structure of foundations and isomorphisms as suggested by the cognitive theory of mathematical metaphor.

Another Vision of Mathematical Cognition

Since we've already started wandering away from the realm of science to the realm of allegory, I'm going to take yet another step away from science, and suggest a new philosophical story about mathematical cognition, which carries the discussion even further away from the world of well ordered hierarchies.

But first, a warning of sorts. It should be obvious that I am strongly influenced by French post-structuralism. This is very explicit in most of my papers, and should be discernible from this book as well. However, I know that working with thinkers like Derrida or Deleuze alienates many readers, and that post-structuralist jargon is impenetrable to the untrained. I decided, for this book, to allow myself only passing references to such thinkers, and almost no jargon. This does, however, come with a price. While I can present much of my thinking in more mainstream language, some of the edge is lost in translation.

The following section, therefore, will be an exception. I will follow here closely some ideas of Gilles Deleuze. I will try to remain accessible to a wide readership in this section too, but if you're allergic to thinkers like Deleuze and the associated style of reasoning, perhaps it's best for you to just skip it. On the other hand, as some mainstream philosophers embedded firmly in the analytic and pragmatic traditions have recently been flirting with Hegel, Heidegger, and Levinas, perhaps negotiating with Deleuze will not come as such a terrible shock.

The book that I follow here is one of Deleuze's less-known works, a rather marginalized interpretation of the work of the painter Francis

Bacon, subtitled *The Logic of Sensation* (2003). I will appropriate and decontextualize this book here to suggest a way of thinking about reasoning with symbols, or, more specifically, with the diagrams of classical Greek geometry.

This interpretation will situate classical geometric practice along a continuum ranging from rigorous optico-linguistic codes to materially constrained embodied gestures set against a noisy background (for the productive relation between reason and noise, see also Michel Serre's [2007] *The Parasite*). What binds this continuum together is what Deleuze calls "haptic vision."

From Diagrams to Haptic Vision

I will begin with some observations by Reviel Netz concerning Greek mathematical practice. Netz originally embedded his deep and insightful analysis in a modular cognitive framework, but I believe it can work even better with the very different interpretation I suggest later (see Latour's critique [2008] for an argument explaining the importance of Netz's work).

According to Netz, a geometric diagram is "a *finite* system of relations.... It is limited in space; and it is discrete. Each geometrical proposition refers to an infinite continuous set of points. Yet ... the lettering of the diagram ... turns it into a system of intersections, into a finite manageable system" (Netz 1999, 34–35). This *finite* system, accompanied by the highly formulaic and restricted Greek mathematical language was "produced from a few simple building blocks, [and allowed] the simplification of the universe." It thus contributed to making the "inspection of the entire universe possible" (Netz 1999, 158, 266).

Netz's description of the coded world of geometric diagrams may recall Paul Sérusier's presentation of abstract painting as "reducing all forms to the smallest number of forms of which we are capable of thinking—straight lines, some angles, arcs of the circle" (quoted by Deleuze 2003, 92). This is what Deleuze calls " 'digital,' not in direct reference to the hand, but in reference to the basic units of a code" (Deleuze 2003, 92). But Deleuze explains that even the units of abstract painting are "aesthetic and not mathematic, inasmuch as they have completely internalized the manual movement that produces them"

(Deleuze 2003, 92). I believe, however, that the internalization of manual movement is true for geometric diagrams as well.

To understand this claim, I note a tension between how diagrams are drawn in principle and in practice. Netz argues that in proposition I.2 of Euclid's *Elements*,

> when an equilateral triangle is constructed in the course of the proposition, one is faced with a dilemma. Either one assumes that the two auxiliary circles [required for the construction] have been constructed as well—but how many steps further can this be carried, as one goes on to ever more complex constructions? Or, alternatively, one must conclude that the so-called equilateral triangle of the diagram is fake. Thus the equilateral triangle of Proposition I.2 [ABD in figure 5.2] is a token gesture, a make-believe. It acknowledges the shadow of a possible construction without actually performing it. (Netz 1999, 54)

The geometric diagram therefore includes a double gesture: the "in principle" gesture of coded rigorous ruler and compass construction, and the practical imprecise drawing. The movement is "digitally" codified in principle (and, where complex diagrams are concerned, only in principle), but "manual" in practice. Note that this tension is reflected in contemporary practice by the in-principle-formalizability of mathematical arguments, which does not necessarily entail formalization in practice.

This manual rendering of diagrams would become even more imprecise with the means available to the Greeks: sand, ashes, or tablets that require advance preparation and so must be used sparingly. Plutarch reports that Archimedes even "used to draw the figures on his belly with the scraper" when his servants rubbed him down with oil (quoted in Unguru and Rowe 1982, 5; this may sound highly fanciful, but I've witnessed and experienced occurrences not very far removed, such as enthusiastic students drawing or writing on their forearms, when paper was not readily available). Of course, at some point the diagram begins to smudge, and attempts at correction end up making an even bigger mess.

In this kind of situation, the following may occur: "one starts with a figurative form, a diagram intervenes and scrambles it, and a form of completely different nature emerges from the diagram" (Deleuze 2003,

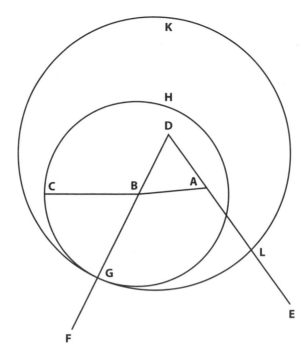

Figure 5.2: *Elements*, Proposition I.2: "To place at a given point (as an extremity) a straight line equal to a given straight line." All quotations from the *Elements* follow Heath's translation.

125). This latter quotation refers to Francis Bacon's practice of painting, but can be appropriated as a description of drawing geometric diagrams as well. To do that, we should read Deleuze's "figurative form" as "geometric diagram" and Deleuze's "diagram" as the messy accidents of material drawing: marks and traits "which are irrational, involuntary, accidental, free, random. They are nonrepresentative, non-illustrative, nonnarrative. They are no longer either significant or signifiers: they are a-signifying traits" (Deleuze 2003, 82). We attempt to draw a geometric diagram, but the manual movement of drawing introduces irrational noise: lines may accidentally superimpose; erasure marks may seem like lines; an accidental stroke of the pen suggests a shape that doesn't belong there; the motion of the belly smudges the diagram that Archimedes draws on it with a scraper; a soldier suddenly appears and disturbs Archimedes' circles ... eventually, the diagram must be discarded and redrawn. And redrawn, and redrawn, and redrawn.

If we are interested in a neat modular narrative, all this noisy mess must be set aside. After we're done redrawing, we come to the scene of writing, where a proof is formally recorded: "the most common [practice] was to draw a diagram, to letter it, accompanied by an oral dress rehearsal—an internal monologue, perhaps—corresponding to the main outline of the argument; and then to proceed to write down the proposition as we have it" (Netz 1999, 86).

Within this neat modular description, however, there is a black box. This box is the internal monologue, which is a rather vague experience. To figure out what's going on in this black box, we can ask, with mathematician Jacques Hadamard, the following question: "what internal or mental images, what kind of 'internal word' mathematicians make use of; whether they are motor, auditory, visual, or mixed"? Einstein was kind enough to reply:

> The psychical entities which seem to serve as elements in thought are certain signs and more or less clear images which can be "voluntarily" reproduced and combined ... this combinatory play seems to be the essential feature in productive thought—before there is any connection with logical construction in words or other kinds of signs which can be communicated to others.... The above mentioned elements are, in my case, of visual and some of muscular type. Conventional words or other signs have to be sought for laboriously only in a secondary stage, when the mentioned associative play is sufficiently established and can be reproduced at will. (Hadamard 1973, 147–48)

Note the muscular component of this practice, which Hadamard and Einstein were so perceptive as to include. The fact of drawing, with one's arm (using a pen), with one's entire body (using a stick on sand or with a scraper on one's belly), integrates the body into mathematical practice. As assumed by the theory of embodied cognition, it is no longer an abstract mental act, but a mix of physical simulation and action. Into the practice of drawing a geometric diagram irrupt "expressive movements, paralinguistic signs, breaths and screams [or, to be less dramatic, sighs of frustration] and so on" (Deleuze 2003, 93).

In a paragraph remarkably in line with Freeman's emphasis on the role of context and attunement in cognitive processes, Deleuze notes that:

> It is a mistake to think that the painter [or the geometer drawing diagrams] works on a white surface.... If the painter were before a white surface, he could reproduce on it an external object functioning as model, but such is not the case. The painter has many things in his head, or around him, in his studio. Now everything he has around him is already in the canvas, more or less virtually, more or less actually, before he begins his work. (Deleuze 2003, 71)

So it is not only the material contingencies of drawing that disrupt a rational coding of Greek diagrams, but also this background and internal noise. Even if the disruptive marks do not appear in the drawn diagram, they are in the mathematician's head.

When a geometric diagram is disrupted, it may look as if "a *catastrophe* overcame the canvas" (Deleuze 2003, 82; Netz [1999, 15] also testifies that his attempts at drawing on sand or ashes were "unmitigated disasters"). This is often a moment of confusion and frustration, but also a moment of detachment. Deleuze's narrative suggests that in such moments, where the coded, abstract space is disrupted, the drawing may break away from its confinement to the material surface of inscription, and be absorbed into the embodied mind through the hand and eye (or the belly, in Archimedes' case). This happens because a gap forms between what the drawing *is* and what it is *supposed to be*. The hand and eye handle and see one thing, but intend (or experience preafference for) something else. The missing object of intention becomes ideal—materially absent but still guiding thought (somewhat like the analytic a posteriori, or the Black-Scholes vignette). In Freeman's terms, we may think of different stimuli (different more or less successful diagrams) that trigger the dynamic neural system to converge to the same attractor (stable activity pattern), associated with the same implications—a neural manifestation of the ideal intention.

But Deleuze does warn us that if we let "catastrophe" run wild, we get no visual space at all. The drawn space would then become a space of action painting (for example, Jackson Pollock): a manual and physical space relating irrational embodied motions. To put it in neural terms: if there's too much noise, the neural system will not be able to form stable attractors.

To avoid this danger, Deleuze explains, the "catastrophe" must not be allowed to "eat away at the entire painting; it must remain limited in space and time. It must remain operative and controlled." Something should "emerge from the catastrophe" (Deleuze 2003, 89). We must now explain how the highly regulated visual-textual products of Greek geometry can emerge from the noisy disruption of drawing diagrams.

If the abstract, coded diagram is broken, but we wish to avoid the wild space of action painting, then the noisy "operative set of asignifying and nonrepresentative lines and zones, line-strokes and color-patches," has to be " 'suggestive' … to introduce 'possibilities of fact' " (Deleuze 2003, 82–83). Bacon explains:

> the marks are made, and you survey the thing like you would a sort of graph. And you see within this graph the possibilities of all types of fact being painted.… [I]f you think of a portrait, you maybe have to put the mouth somewhere, but you suddenly see through the graph that the mouth could go right across the face (Deleuze 2003, 160).

In a badly drawn geometric diagram, a crooked circular arc may suggest using a parabola to solve a problem; a line drawn out of place may suggest a useful auxiliary construction; a soldier disrupting one's circles may suggest integrating him into the diagram as the axis of a cone, embedding the diagram in three dimensions to figure things out … In Freeman's terms, these accidental interruptions may push the network to an unexpected attractor that can suggest a new action, out of the ordinary.

So far we gained two things from the "catastrophes" of material drawing. First, while seeing and drawing one thing, we can envision and hold on to another intention. Second, disruptive noise can sometimes be integrated into a new structure. Since these disruptions are not only on the canvas but also "in the head," they may include other forms of reasoning that the mathematician is engaged with, triggering metaphorical transfers of knowledge.

Once something emerges from a disrupted geometric diagram, we have before us that which Deleuze calls "Figure": "a shifting sequence or series (and not simply a term in a series); it is each sensation that exists at diverse levels, in different orders, or in different domains" (Deleuze 2003, 33). What are these "levels" in our case? One level consists of seeing in a diagram a sequence of previous diagrams that have been discarded. Another is seeing in a diagram not what is actually drawn, but what we intended to draw. Yet another includes the integration of drawing "catastrophes" into new ways of processing the problem.

In neural terms, the first level can be associated with the impact of past drawing experiences on the neural network. The second level can be related to preafference: directing sensory experience toward whatever it is that action prepares us to sense. The third level can be related to the formation of new attractors or convergence to unexpected existing attractors due to noisy stimuli. A noisy diagram may trigger neural networks to go through a sequence of shifting attractors in brain areas that handle sensing (vision) and acting (drawing), turning the static image into a multi dimensional and dynamic experience. Given this dynamic, the diagram is no longer a static object, but a trigger for a dynamic chain of interpretations. Diagrams can then be read as telling a story or narrating a proof that may later be formalizable in a language that can be consensually evaluated.

Deleuze refers to the embodied-cognitive capacity to handle this kind of multilevel diagrams by the term "haptic vision.... It is as if the duality of the tactile and the optical were surpassed" (Deleuze 2003, 129; "haptic" refers to the sense of touch). Initially, the hand only drew, and the eye only observed the result. But the emergent Figure involves a *mani*pulation or *hand*ling without resorting back to the hand, using only the associated enriched mode of *vision*. Indeed, we see in the diagram a sequence of past drawings, intended drawings and possible integration of noise into a new drawing—all without any actual manual redrawing. We can therefore attribute to the eye some of the former powers of the hand, resulting in a haptic enhancement of vision. Something similar happened when abbacus masters looked at a number and saw other possible numbers (chapter 2), and when contemporary mathematicians look at an *x* and see various superposing semantic and syntactic interpretations (first section in chapter 4).

Haptic Vision in Practice

Let's try to pour some concrete mathematical content into what I tried to subsume under Deleuze's concept of haptic vision.

First, let's look at the language of classical Greek texts. These texts commonly order readers to "let a circle have been drawn," or "let the point A have been taken" (Netz 1999, 51, 25). In Greek, these are perfect imperatives, orders to have already completed an action. To whom can such an order be addressed? If we do not dismiss these commands as rhetorical curiosities, then they are directed at something that, like the hand, has the power to make things happen, but, unlike the hand, can also sense them as already-having-happened. Haptic vision has that power because it integrates the neural sediments of manual motion and a history of discarded drawings. I use the term "neural sediments" rather than "memory," as it is not about reactivating something safely stored for opportune retrieval; rather, it is about the way past experiences formed and reformed our cognitive infrastructure and the stable patterns (attractors) to which this infrastructure can give rise. Haptic vision thus has the capacity to have had drawn and lettered, to reactivate the past without drawing on a memory—that is, to sense in an actual Figure both a newly given present and a completed past.

At this point, it should no longer be odd that we can "Let the points A, etc. be imagined as the points of the angles of the inscribed pentagon," even when this polygon is not actually drawn in the diagram of *Elements* Proposition IV.12 (figure 5.3). The cumulative experiences of erasing and redrawing endow haptic vision with preafference, that is with the capacity to sense lines and points without actually resorting back to the drawing hand.

The same experience enables haptic vision to retrace elements that are no longer mentioned in either diagram or text. In one of the cases that Netz analyzes (cutting a sphere in a given proportion; Netz 2004, 54), the readers "imagine a watermark underneath Diocles' diagram [that] has a sphere ... and two cones," even though Diocles actually discarded the sphere and the cones that had motivated the analysis from his diagram and text (figure 5.4). The cones and sphere are part of what's there "in [the geometer's] head or around him," as one of the

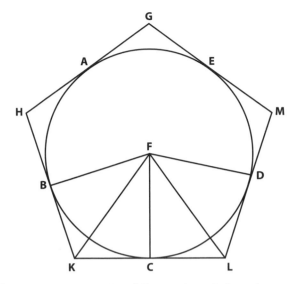

Figure 5.3: *Elements*, Proposition IV.12: "About a given circle to circumscribe an equilateral and equiangular pentagon."

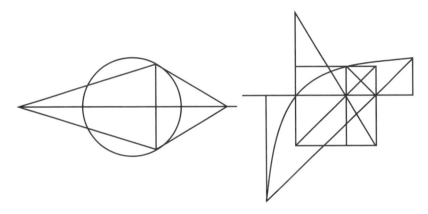

Figure 5.4: Archimedes' diagram (left) presenting the problem of cutting the sphere in a given proportion, and Diocles' diagram (right) that suppresses the circle and cones. Based on Netz (2004), 12, 42.

most likely attractors of the cognitive dynamics in the given context, or as an operator constraining this dynamic.

This sensing as completed of that which is just now being given, of that which is not actually drawn and of that which has been discarded is not static. Suppose we are ordered to "Let *some* point be taken on

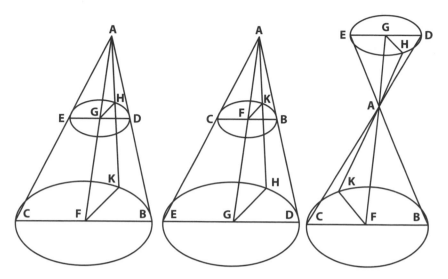

Figure 5.5: Apollonius's cones. Based on Fried and Unguru (2001).

the circle, *A*" (Netz 1999, 22), or "Let a chance point be taken on *AB*" (Netz 2004, 77). Here we are ordered to sense a point as general or as random. To do that we require more than just an observing eye or a drawing hand. We must sense with the given point the possibility of alternative points being taken elsewhere. Since haptic vision emerged from the random marks and "catastrophes" of discarded diagrams, it retains the neural sediments of points marked elsewhere. As the experience of any such point may direct the brain dynamic to the same attractor, each point could be considered as equivalent to the others. This neural dynamics may be the material cognitive counterparts of the general, random and possible.

Sometimes geometric elements are not only general or random, but actually mobile. Fried and Unguru follow Apollonius in stating that a "straight line generating [some conic] surface is moved" to generate the cone, even though the diagram is still. If indeed the "three diagrams given in the body of the proof … may be thought of as three 'snapshots' of the generation of the surface just described" (Fried and Unguru 2001, 69–70; figure 5.5), it is because the haptic eye that watches over them can see, as Deleuze put it, "a shifting sequence or series" (Deleuze 2003, 33). The diagrams may be "indistinguishable

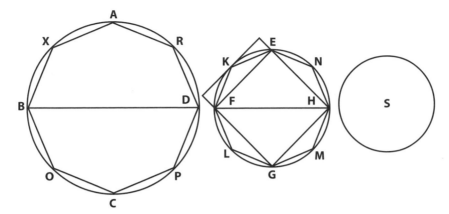

Figure 5.6: *Elements*, Proposition XII.2: "Circles are to one another as the squares on the diameters."

except for the labeling," but "one can *see* in the proof and in the diagram the genesis" of the relevant geometric objects. The diagrams of Apollonius urge us "to imagine a cone being increased from one base to another, or, rather, a sequence of cones with ever-increasing bases" (Fried and Unguru 2001, 69–70). From the point of view of the haptic eye, this wouldn't be imagining; it would be sensing in the Figure a host of past discarded sketches as a shifting sequence of possibilities or a series of consecutive brain dynamic attractors, each triggered by its predecessor. Haptic vision includes, together with the dimensions of present, past, absent, discarded, random, general, and possible also a dimension of motion.

Ian Mueller quotes Heath quoting "Simplicius' report according to which the fifth-century sophist Antiphon claimed that a polygon of sufficiently many sides inscribed in a circle would exhaust the circle" (Mueller 1981, 234). This is precisely the kind of perception that thick-lined material drawing and its "catastrophes" may conjure. Indeed, a circle traced manually by a thick pencil may be indistinguishable from a many-sided polygon. The haptic eye unfolds this contradiction into a "shifting sequence or series" of possible stages—in our case, a series of polygons that serve as improving approximations for the circle. This maneuver is integrated into a rigorous form of reasoning in exhaustion arguments such as that of *Elements* XII.2 (figure 5.6), where

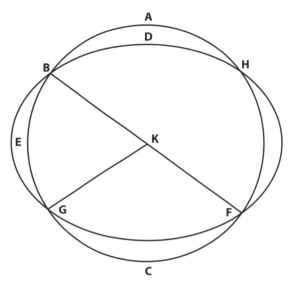

Figure 5.7: *Elements*, Proposition III.10: "A circle does not cut a circle at more points than two."

one proves a theorem about the ratio of circles by resorting to a sequence of polygons that increasingly approximate the circles.

Finally, haptic vision can sense not only the past, general, absent, mobile, and sequential, but even the absurd. In Proposition III.10 of the *Elements* (figure 5.7), two circles are drawn intersecting at four points in order to prove the absurdity of this situation. The sedimented experience of "catastrophic," messy sketches of crooked circles can allow haptic vision to hold onto the absurd diagram, and still obtain a rigorous geometric proof. Indeed, haptic vision is used to the gap between diagrams and what they purport to draw. Under haptic vision, the diagram is not just what is seen or drawn, but also something that can be saved from "catastrophic" drawings to form a reasoned Figure. Under specific intentions, Haptic vision senses by means of preafference something other than what is present, and breaks down the absurd diagram into elements that make sense.

In order to make rational sense, the Figures that emerge from drawn diagrams are stratified into orders of past, present, absence, erasure, chance, generality, possibility, motion, and counterfactuality. Haptic vision operates not by subjecting Figures to a single abstract code, but by an evolving cognitive dynamic (rather than a fixed cognitive state) that articulates and communicates the preceding possibly contradictory orders. Each order may be considered as arising from the same stimulus (the diagram) feeding into a given brain area, but under a somewhat different global operator (different expectations, states of mind, or intentions); alternatively (and probably more in line with Freeman's thought) each order may be considered as the result of different stimuli feeding into a given brain area under the global operator, or symbol, that the diagram presents. It is the haptic eye's articulation and recomposition of orders of sensation into a serial development or "movement in all its continuity" that regulates what Deleuze calls the "logic of sensation" (Deleuze 2003, 35). This is the logic that turns noise into creative input, and binds together conflicting constraints and superposing interpretations into a system of layers that are piecemeal formaliz*able*, but not *actually* subject to any global formalism. This logic is what the various case studies and arguments of this book attempt to illustrate.

The economy of this haptic vision is complex. We often think of the gaze as having power over its object. At the same time, we think of the gaze as being captured by objects. Haptic vision introduces a sense of balance into this economy. The Figure compels itself on us, but we have the power to manipulate it by means of haptic vision. Indeed, the haptic gaze invests the Figure with a compelling dynamism. This "logic of sensation" is perhaps one way to articulate mathematical truth. Truth emerges as a balance between our power to manipulate an object while being compelled by its law.

CHAPTER 6

■ ■ ■ ■ ■ ■ ■ ■

Mathematical Metaphors Gone Wild

IN THIS CHAPTER, I WILL REFLECT on two case studies. The first will consider four medieval and early modern examples of relating algebra to geometry. These examples will show that when two mathematical domains are linked, what passes between them cannot be reduced to "inferences," as assumed by the theory of mathematical metaphor. The second case study will review notions of infinity since early modernity. The purpose of the review will be to show that these notions are far too variegated and complex to be subsumed under a single metaphor—namely, Lakoff and Núñez's basic metaphor of infinity, which tries to read all mathematical infinities as metaphorically projecting final destinations on indefinite sequences.

What Passes between Algebra and Geometry

We'll consider here four historical case studies related to the algebraization of geometry. For each of these case studies I will ask, "What is it that's transferred between geometry and algebra?" If in each case we have a transfer of entities and inferences from one conceptual domain to another, then our findings support the theory of Lakoff and Núñez (2000). If, however, we can't reduce whatever is transferred between algebra and geometry to entities and inferences, then mathematical metaphor must be something more complex.

The analysis refers to the specific case studies presented, and does not necessarily reflect the overall approach of the relevant mathematicians. I shuffled the chronological order of the examples in favor of the logical buildup of the analysis. Wherever I used anachronistic notations, I tried to use them in a way that won't tamper with my analysis here.

Piero della Francesca (Italy, Fifteenth Century)

The problem: There's a circle with diameter 12, we want to inscribe a[n isosceles] triangle such that one of its sides is 8. I ask for the other two sides (della Francesca 1970, 211).

Solution (summary with anachronistic notation): We take the altitude of the triangle as the unknown thing (anachronistically denoted x). Since chords of a circle intersect proportionally, and since the altitude bisects the base, we have

$$x(12 - x) = (8/2)^2.$$

Rearranging the equation, we get

$$x^2 + 16 = 12x.$$

Solving according to the standard rule, we obtain

$$x = 6 + \sqrt{20}$$

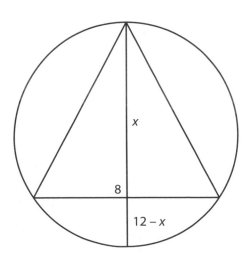

Figure 6.1: Piero's diagram.

According to the Pythagorean theorem, the side of the triangle is

$$\sqrt{4^2 + (6 + \sqrt{20})^2} = \sqrt{72 + \sqrt{2880}}.$$

Analysis: We have here a combination of geometric inferences (circle proportion theory, the Pythagorean theorem) and algebraic inferences (forming, simplifying, and solving quadratic equations). We also have here a nonclassical conflation of lines and numbers, which are treated as interchangeable entities, in violation of the standards of classical Greek geometry.

We can easily describe this case study in terms of a conceptual blend (lines, numbers, and algebraic unknowns are correlated) and a conceptual metaphor (algebraic inferences are imported to geometry). This fits well with Lakoff and Núñez's theory of conceptual metaphor. Here, *algebraic entities and inferences are transferred into geometry*. Note, however, that the direction of transfer violates the requirement of transferring inferences from more concrete domains to more abstract ones ...

Omar Khayyam (Central Asia, Eleventh Century)

The problem: A cube equals sides and a number (Wöpcke 1851, 32). Anachronistically, this reads: $x^3 = ax + b$.

Solution (summary with some anachronistic notation): Let the line AB be the side of a square equal in area to a, the coefficient of the side x in the problem (that is, $AB = \sqrt{a}$). Let the line BC be the side of a box that is equal in volume to the number in the problem (b), and whose base is the square of area a built on AB (that is, $BC = b/a$). Let DBE be a parabola with parameter AB, and ZBE a hyperbola with parameter BC. E is their point of intersection, and EH and ET are perpendicular to the continuations of CB and AB, respectively.

From the properties of conic sections, it follows (denoting "x to y is as z to w" by "$x{:}y :: z{:}w$")

$$AB{:}BH :: BH{:}BT :: BT{:}CH.$$

This means that

$$sq(AB){:}sq(BH) :: BH{:}CH.$$

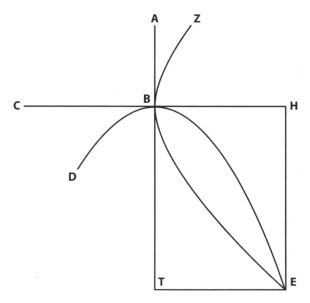

Figure 6.2: Khayyam's diagram.

By cross multiplication, we have

$$\text{cube}(BH) = \text{box}(sq(AB),\ CH) = \text{box}(sq(AB),\ BH) + \text{box}(sq(AB),\ BC).$$

Given our choices of AB ($\sqrt{a}$) and BC (b/a), we find that BH is the required side of the cube (x).

Analysis: The problem is explicitly formulated in terms of the rhetorical Arabic *Shay-Mal* algebra, and is part of a treatise that systematically covers a long list of algebraic problems (cubic equations). Khayyam's conception of algebra, however, is as follows: "The art of algebra … has as its goal to determine unknowns, either numerical or geometrical" (Wöpcke 1851, 1). This means that algebra is a language that is used to speak about arithmetic and geometry. However, in Khayyam's book, arithmetic interpretations of algebra are limited to the simplest equations (for example, 16–18), and are quickly set aside.

When the algebraic problems are interpreted geometrically, the reasoning is highly classical, relying strictly on geometric inferences (except for the conflation of numbers and lines; see Netz 2004 for an analysis). Moreover, in the solutions, division into cases is based on geometric, rather than algebraic considerations, and even numerical examples undergo a highly geometric analysis.

While we can speak here in terms of an underlying arithmetic-geometric blend that conflates numbers and lines, we can't say that algebraic inferences are transferred into geometry—all inferences in Khayyam's work are geometric. Nevertheless, the role of algebra here is significant. Algebra here is the means to provide a new organization of geometric knowledge. The same arguments that served Archimedes for the geometric task of slicing a sphere in a given proportion are reset in Khayyam's work inside a system of problems that depends on algebraic forms of expression. Algebra furnishes Khayyam with a system of problems, rather than inferences or entities. *What is transferred here from algebra to geometry is an organization of knowledge*—not entities, hypotheses, and inferences, but constitutive problems.

René Descartes (France, Seventeenth Century)

The problem: Solve the quartic equation $z^4 = pz^2 + qz$ (Descartes 1954, 196).

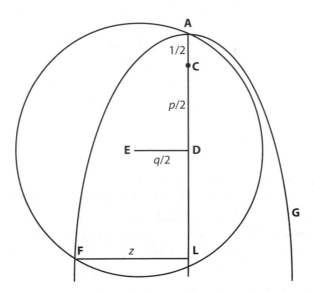

Figure 6.3: Descartes's diagram.

Solution (summary with very little anachronism): Let FAG be a parabola with parameter 1. Draw along the parabola's axis AC and CD of length $1/2$ and $p/2$, respectively. Then draw a perpendicular DE of length $q/2$. Draw around E a circle of radius AE. Let F be the intersection of the parabola and the circle, and draw FL, a perpendicular to the axis of the parabola. Setting $z = FL$ solves the equation.

Indeed, by the property of the parabola

$$AL = FL^2 = z^2.$$

Since F is on the circle,

$$FE^2 = AE^2 = (q/2)^2 + ((p + 1)/2)^2.$$

Also, according to the Pythagorean theorem,

$$FE^2 = (z - q/2)^2 + (z^2 - (p + 1)/2)^2.$$

Comparing the two values of FE and simplifying, we obtain that z indeed satisfies the original equation.

Analysis: The book's title, *La Géometrie*, reflects its motivation. Indeed, the first two parts of the book deal with algebraically assisted solutions of geometric problems from the classical corpus. In that sense, the approach is comparable to that of Piero della Francesca (though much more sophisticated). But the third part of the book, from which the problem is taken, opens with purely algebraic problems and a purely algebraic analysis: cubic and quartic equations are classified and simplified by algebraic means. Much effort is spent on transforming irrational coefficients into rational ones and on deriving rational solutions—procedures that are clearly not geometric. In fact, it is only half way into the third part of the book that a diagram appears.

Moreover, unlike the situation in Khayyam's case, the analysis of the geometric construction (and most likely, the motivation too) is mostly algebraic. The details of the solution make it highly plausible that it was derived by transforming the original equation to a comparison of completed squares, and then providing a geometric interpretation (see Bos 2001). Indeed, the parabola-circle intersection is a geometric representation of the algebraically derived equality between the two algebraic expressions $(z - q/2)^2 + (z^2 - (p + 1)/2)^2$ and $(q/2)^2 +$

$((p+1)/2)^2$. The use of a geometric representation to solve an algebraic problem is the novelty introduced here by Descartes.

Note, however, that Descartes knew the Italian arithmetic formula for solving the preceding equation and quoted it explicitly. What's the point, then, of Descartes's geometric representation? The answer lies in the next pages of Descartes's book, where equations of degrees 5 and 6, which are not solvable by algebraic means, are solved by similar geometric representations (depending on more advanced curves). The geometric representation of algebraic equations therefore serves to enhance our ability to solve algebraic problems. Note how radical the impact of geometric representation is: equating the two *algebraic* expressions for the length of *FE* counts as the statement of a problem; but the equivalent *geometric* drawing of the intersecting curves counts, according to Descartes's contemporary standards, as the representation of a solution.

This solution obviously depends on an underlying blend of numbers, unknowns, and lines, but it cannot be reduced to a transfer of entities or inferences between algebra and geometry. It's not a geometric inference that helps us solve the algebraic problem, but geometric forms of solution representation. The novelty here is *a transfer of means of representation from geometry to algebra.*

Rafael Bombelli (Italy, Sixteenth Century)

In his manuscript book III from the 1550s (Bologna, Biblioteca dell'Archiginnasio, B.1569), Bombelli presented a host of recreational/commercial mathematical problems, and solved them algebraically. In Book IV, some of these problems are treated again, this time geometrically. Let's take a look at one such problem.

The problem: Algebraically, we are concerned with a partnership problem (fol. 134r). To put it in slightly anachronistic terms, three partners contribute x, $2x + 4$, and $x(2x + 4)$, respectively. Their business made 300, and when they divided the profits, the first partner got 20.

Solution: The statement of the problem yields the relation

$$x : (x + (2x + 4) + x(2x + 4)) :: 20 : 300,$$

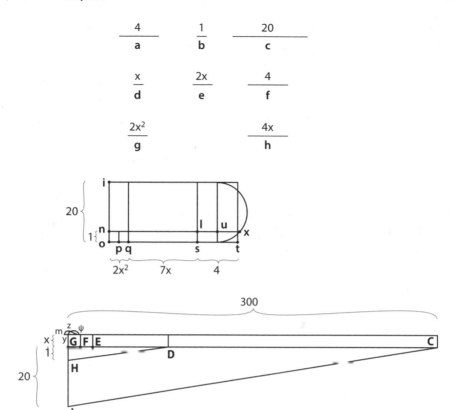

Figure 6.4: Bombelli's diagram.

which by cross multiplication translates into

$$20(2x^2 + 7x + 4) = 300x.$$

In the geometric version (Bombelli 1929, §102), this equality translates into an equality between the rectangle *iot* (where *io* is 20, *oq* is $2x^2$, *qs* is 7x, and *st* is 4) and the rectangle *yGC* (where *yG* is x and *GC* is 300). Note here that the three parts of the line *ot* involve different powers of *x*—we will explain how Bombelli handles this issue after we describe the steps of the solution.

The first algebraic step is to divide the equation by 20, which yields

$$2x^2 + 7x + 4 = 15x.$$

This step is reflected geometrically by rescaling the two rectangles: the top rectangle is rescaled along the vertical dimension according to the ratio *on:oi* (1:20), and the bottom rectangle along the horizontal dimension according to the same ratio, this time represented as *GH:GI*, which equals *GD:GC*.

The next algebraic steps are removing $7x$ from both sides and dividing by 2 to obtain

$$x^2 + 2 = 4x.$$

Geometrically, this is represented by removing from the top rectangle the rectangle with base *qs* and from the bottom rectangle the rectangle with base *ED*, and then halving the remaining rectangles along the horizontal dimension, yielding an equality between the sum of *nop* and *uls* on the one hand and *yGF* on the other. Finally, the algebraic equation is solved by the standard formula that yields $2 + \sqrt{2}$, which is translated into a geometric construction of the line ψm (we skip the details of this last construction).

The crucial point here is that in this diagram, there are three regimes of lines: known, multiples of the unknown *x*, and multiples of squares of the unknown *x*. Since a priori we do not know the value of the unknown, we do not know how to represent such lines as *qs* and *st*. Nevertheless, Bombelli has to draw them somehow.

To do that, the unknown line *d* (representing *x*) is drawn as equal to the unit line *b* (representing 1). This way a line that is a multiple of the unknown line (say, *qs*, representing $7x$) is equal *in length* to the coefficient of the unknown (7). So the length of a line in this diagram can represent different things depending on whether the line is known or unknown: the line's magnitude or the coefficient of the unknown, respectively.

A line here is more than just a representation of its length—it has a length and a register (known or unknown). One could perhaps think of this situation as a new conceptual blend of algebra and geometry, but this would violate Lakoff and Núñez's requirement of "two distinct cognitive structures with fixed correspondences between them." It is precisely because the correspondence here is *mixed* (one correspondence for the known lines and a different kind of correspondence for unknown lines) that this representation works (note that x is not a variable, but an unknown number determined by the problem).

Then there's the issue of representing the line whose length is $2x^2$. Here Bombelli takes the rectangle formed by the lines representing $2x$ and x, and constructs another rectangle of equal area on the line b of length 1. The length of the other side of the new rectangle is used to represent $2x^2$. Now, since the unknown line is represented by a line of length 1, the line representing $2x^2$ is actually of length 2. A careful reading of Bombelli's language shows that the lines representing the square of the unknown often have an extra implicit dimension (the other side of the rectangle used to construct them), so these lines are sometimes understood as rectangles whose other side, of length 1, belongs to a third implicit dimension.

Analysis: We see here a rather intricate form of representation, which requires several levels of correspondence. Each line is read according to its *length* and according to its *register* (known, unknown, or square), and the diagram is sometimes understood as two dimensional and sometimes as having a third implicit dimension. This blend of algebra and geometry is not an instance of Lakoff and Núñez's "fixed correspondence." Mixed correspondences in diagrams are not strictly an early modern phenomenon (see, for example, the analysis of modern Dynkin diagrams by Lefebvre 2002). Such phenomena are better understood by the notion of superposing interpretations or the multidimensional vision of Deleuze's haptic eye.

But we still haven't answered the question of what exactly is transferred here from one conceptual domain to the other. We might say that, as with Descartes (and anticipating some of his techniques), we import a geometric form of representing solutions into algebra. But in the case of Descartes, this transfer of means of representation enabled solving equations that could not be solved otherwise, whereas here, and throughout Bombelli's work, no such thing occurs. Bombelli's geometric representations are complicated and creative means of retracing algebraic calculations step-by-step with lines rather than algebraic and arithmetic signs.

Bombelli's motivation for his apparently useless representation can be recovered from the following quotation:

I had in mind to verify with geometrical demonstrations the working out of all these Arithmetical problems, knowing that these two sciences (that is, Arithmetic and Geometry) have between them such accord that the former is the verification [*prova*] of the latter and the latter is the demonstration [*dimostration*] of the former. (Bombelli 1966, 476)

What is transferred here between geometry and algebra is not just means of representation, organization of knowledge, entities, or inferences. *What is transferred here from geometry to algebra is epistemological status.* It is because these two domains can be *made* equivalent (by a nonfixed, multilayered correspondence), that the young science of algebra becomes epistemologically more robust. The purpose is to provide a mechanism of geometric formalization, which would help algebra meet Bombelli's contemporary validity constraints. This is a practice of providing relevant interpretations in order to establish mathematical truths.

Conclusion

The preceding analysis demonstrates two facts. First, it's not only entities and inferences that are transferred between mathematical domains. The transfer of organization of knowledge, means of representation, and epistemological status is crucial for mathematics, and the reader is invited to think of other cognitive elements transferred between mathematical domains.

Such transfers are not special to the preceding examples, and take place in contemporary mathematics as well. Reframing mathematical structures under new schemes (say, the theory of functions under functional analysis) brings about new systems of problems, even if they may then be solved by classical means; transferring means of representation (for example, reinterpreting a differential equation in a richer setting that allows for "weak" or "generalized" solutions, or solving problems "in high probability" by random constructions) brings new representations of solutions; and the epistemological status of mathematical structures can be validated by importing models and techniques from "foreign" domains (for example, model-based consistency proofs and nonconstructive topological or set theoretic existence proofs).

Second, the transfer of knowledge between domains may be about *mixed*, rather than *fixed*, correspondences that consider mathematical signs in several ways at the same time, possibly involving underlying interpretive contradictions (the same line simultaneously representing different magnitudes in the same diagram). This layering is in line with a practice of seeing various dimensions or layers in a single diagram, or the notion of the haptic eye.

This practice is, again, not an isolated occurrence. It relates to simultaneously representing (as discussed in the section on interpretation in chapter 3) a matrix as an operator (rotation) and operand (square), or other positions in an open ended functional hierarchy. It relates to the superposition of interpretations in chapters 2 and 4 as well.

The conclusion is that in thinking about mathematical transfer of knowledge between domains, we should renounce two reductions: first, the reduction of mathematics to entities and inferences, and second, the reduction of mathematical formal consistency (the fact that, when properly disambiguated and contextualized, no two mathematical statements contradict each other) to unique and consistent underlying metaphorical (or cognitive) structures.

A Garden of Infinities

In the previous chapter, I tried to argue for a view of mathematical metaphors as a dynamic process involving interrelations between vaguely delimited domains. Here, I am going to demonstrate this point by exploring concepts of infinity. The rich historical variety of infinities will be contrasted with an attempt to ground all mathematical conceptions of infinity in a single cognitive metaphor, Lakoff and Núñez's so-called basic metaphor of infinity, or BMI (2000, 161).

The BMI consists of two steps: first one conceives of continuous processes as iterative processes (that is, breaking them into reiterated minimal steps), and then one projects the existence of a unique final state from the domain of completed iterative processes to that of indefinitely iterated processes. The result of these metaphors are notions of limit (projecting a "final object" on an infinite process) and complete infinity (projecting a "final magnitude" on the infinite sequence of unboundedly increasing or decreasing magnitudes).

For Lakoff and Núñez, BMI is the "single general cognitive mechanism underlying all human conceptualization of infinity in mathematics" (2000, 170). I will show that this claim fails both in the context of limits and in the context of complete infinities.

Limits

According to the BMI framework, limits of functions are conceived as limits of sequences extracted from these functions. In other words, the limit of the function $f(x)$ as x approaches a is conceived as the limit of the sequence $f(x_n)$ as x_n approaches a. This is the first step of the BMI, which reads continuous processes as iterative ones.

Historically, however, this wasn't the case. Indeed, the first canonical instances of limits in early modern European mathematics came from continuous physical processes, and did not depend on discrete reductions. In early calculus, notions corresponding to limits were mostly related to physical motion, with little or no reference to limits of sequences.

But even more importantly, these notions of limit did not necessarily depend on bringing an indefinite process to completion. When thinking with Newton about moving bodies across a finite span of time and space, one considers their velocities. Velocity can be viewed as an average velocity across an interval (space traversed divided by time spent) or as a momentary velocity: the velocity that the body would have, if it continued to move without change (this is indeed William Heytesbury's fourteenth-century definition; see Clagett 1961, 236). For Newton, as a body reaches the limit (namely, end) of its finite motion, its velocity reaches the limit of the finite and continuous span of values through which the body's velocity has gone. This is what Newton speaks of when he discusses "last velocities": "by last velocity I understand that with which the body is moved neither before it arrives at the last position and its motion ceases, nor thereafter, but just when it arrives" (Ewald 1996, I, 60).

So what is it that's infinite about this limit? I quote Newton's explanation (the evanescent quantities referred to here are the evanescent time and space whose ratio may be used to define momentary velocity):

> It may also be maintained that if the last ratios of evanescent quantities [the last velocity] are given, their last magnitudes [those of the evanescent time and space themselves] will also be given; hence all quantities will consist of indivisibilia.... But this objection rests on a false hypothesis. Those last ratios with which quantities vanish are not truly the ratios of last quantities, but limits towards which the ratios of quantities decreasing without limit always approach; and to which they approach nearer than by any given difference, but never exceed, nor attain until the quantities diminish *in infinitum*. (Ewald 1996, I, 60)

The limit here is that of a continuous and finite process and of finite velocities bounded away from zero. The evanescent quantities (diminishing spans of time and space toward the end of motion) indeed decrease without end—that is indefinitely, with no final state, "without limit." Indeed, their disappearance at zero is not considered a finite *state*, as zero is viewed as absence, rather than an existential state (similarly to the sequence of increasing integers having no final state). But the ratios, whose limit is what Newton is after, span a finite and continuous range of values. The limit is simply the end of this range.

The operative metaphor here carries the notion of an end, edge, or limit from the domain of geometric magnitudes to that of ratios such as velocity. Newton indeed states: "And since these limits [of velocities] are certain and definite, to determine them is a purely geometrical problem" (Ewald 1996, I, 60). This geometric metaphor was well entrenched in the relevant intellectual culture. Book V of Euclid's *Elements*, which deals abstractly with ratios of any magnitudes, always draws magnitudes as lines. If we think of velocity as a finite magnitude contained in a finite interval, there's no problem thinking about the limit of the velocity (in the sense of the edge or end of the interval of velocities) of a body reaching the end of its finite and continuous motion.

Newton's underlying metaphor here is: ratios are magnitudes, and finitely bounded magnitudes, just like line segments, have ends or limits. The problem arises if we object that velocity, being a ratio, is not itself a magnitude. This was indeed Berkeley's objection to Newton. He claimed that while intervals and surfaces have limits (namely, their ends and edges), talking about the limit of velocities makes no sense: "A point may be the limit of a line: A line may be the limit of a surface: A moment may terminate time. But how can we conceive a velocity by the help of such limits?" (Ewald 1996, I, 78). Berkeley calls the geometric limit metaphor, and rejects it. According to Berkeley, ratios are not "limitable" geometric magnitudes.

It's important to emphasize that I endorse the claim that Newton did employ some BMI-like constructions (for example, his notion of "moment"). But I argue that Newton's concept of limit is not reducible to BMI, and involves other metaphors as well.

But BMI and "ratios are geometric magnitudes with ends" are not the only metaphors involved in the notion of limit. Other spatial metaphors operate, for example, in the notion of accumulation points of a given set of points. Accumulation points *can* be thought of as the limits of all convergent sequences that can be extracted from a given set (excluding sequences with constant tails), but this view, much like thinking of function limits as derived from sequence limits, misses an important mathematical metaphor. Accumulation points can also be thought of as points that cannot be spatially separated from the given set.

Now, even this revised formulation can still be interpreted as an application of BMI: we could be thinking in terms of a sequence of ever smaller intervals around the accumulation point that fail to separate it from the points of the given set. But we can also think in terms of taking a single *generic* interval around the accumulation point, and showing that it necessarily contains points of the given set. These two maneuvers are, of course, mathematically equivalent; but metaphorically they are distinct.

To make things concrete, think of the accumulation point 0 with respect to the given set $A = \{1/n : n$ is a positive integer$\}$. Working with BMI, we would say that as we take intervals of decreasing radii $(-1,1)$, $(-1/2,1/2)$, $(-1/3,1/3)$, and so on, around zero, we will find inside these intervals, respectively, the points 1/2, 1/3, 1/4, and so on, of the given set A. Therefore, 0 cannot be separated from the set A, and at the imputed "end" of the process, the intervals will shrink to the accumulation point 0. But we may also discard the process interpretation, and say that between 0 and a generic barrier $\varepsilon > 0$ lies the point $1/(|\frac{1}{\varepsilon}| + 1)$ that belongs to A.

These mathematically equivalent statements represent two different metaphors: the first is BMI, but the second thinks of accumulation points as points that cannot be spatially separated by a barrier from a spatially deployed set. There is no *process* here; the metaphor stands perfectly still. Sequentializing this process imposes upon it a metaphor that is not always relevant, and covers over static and geometric interpretations of some limit related concepts.

Different metaphors of limit do not end there. When it comes to infinite sums (series) of functions, a BMI-compatible reconstruction of a series as a sequence of partial sums is not the only available path, and for many mathematicians in the eighteenth century not even the dominant one. First, in some cases, the discrete variable of the sum was read as continuous, reversing the steps postulated by BMI. Moreover, the term "convergent series" was sometimes used to refer to series with diminishing terms, rather than to series whose sequence of partial sums has a limit. One even spoke of series that are convergent up to a point, and then diverge, as in the case of asymptotic series.

While the sum of a series (the limit of its partial sums) was indeed an important notion, the value of a series (as Euler called it in a letter to Goldbach; see Jahnke 2003, 122), that is, the value of a function from which the series can be derived according to some algorithm, also played a major role. The series $1 - x + x^2 - x^3 + \cdots$, for example, was an expansion of $1/(1 + x)$. Therefore, the value of $1 - 1 + 1 - 1 + \cdots$, obtained from substituting 1 for x in the preceding function, was $1/2$.

Today, following Cesàro (building on an interpretation suggested by Leibniz), we may think of this limit as the limit of the averages of increasingly many partial sums—a conception that does fit BMI. But as we mentioned in the previous chapter, it was a different, algebraic way of thinking that dominated at the time. The value of the infinite series depended on algebraic manipulations, and was not reducible to the limit value approached by partial sums.

Finally, one more metaphor, this time from higher modern mathematics. The sequential process envisaged by BMI also fails when considering the limit of a sequence with respect to an "ultrafilter." An ultrafilter on the positive integers is a set of subsets of the positive integers with the following properties:

- The empty set is not a member of the ultrafilter.
- If a set belongs to the ultrafilter, then so does any set that contains it.
- The intersection of two sets that belong to the ultrafilter also belongs to the ultrafilter.
- For any set of integers, either it or its complement belong to the ultrafilter.

Ultrafilters are a difficult notion to swallow. The trivial examples of ultrafilters are those that consist of all sets of integers that contain some fixed number n (the preceding properties are trivially verified). The existence of nontrivial ultrafilters is independent of the ZF axiom system, and follows from the axiom of choice.

Now the definition of a limit with respect to an ultrafilter and the standard Weierstrassian definition can be subsumed under the following template: L is the limit of a sequence $\{x_n\}$ if for every positive ε, the members of some "large" subsequence $\{x_{n_k}\}$ are all within distance ε of L. In the Weierstrassian definition, a "large" subsequence is some tail of the original sequence (a set of the form $\{n \in \mathbb{N} \mid n > n_0\}$). In the ultrafilter-based definition, a "large" subsequence is a subsequence whose set of indices belongs to the ultrafilter.

A limit with respect to an ultrafilter no longer depends on a BMI-like step-by-step process, because unlike the tails of the integers, the sets that belong an ultrafilter are not linearly ordered, and the limit of a sequence does not represent its "end" state, but rather the value best approximating the "bulk" of the sequence in some sense.

One could say that the ultrafilter limit takes subsequences that are considered "important" in some sense (their sets of indices belong to the ultrafilter), and then progressively extracts important subsequences, in line with a BMI limit, so as to obtain improving approximations of a fixed limit value. But this approach has little to do with the way the notion of limit with respect to an ultrafilter is motivated and used in actual practice, which usually carries a more logical or abstract-algebraic flavor.

The point is that while we can, with some effort, reinterpret many limit notions in terms of BMI, it can't be assumed to be their underlying generative mechanism. While we can work with reconstructed BMI interpretations, we also actually work with other superimposed and deferred interpretations. Trying to reduce all of them to BMI does not reflect mathematical practice any better than expressing them in some fixed formalist framework.

Infinitesimals and Actual Infinities

Given the plurality of metaphors that play a part in the notion of limit, it's no surprise to find that BMI provides only a very partial account of actual infinities and infinitesimals.

Lakoff and Núñez propose two applications of BMI to represent what they consider the Euclidean and modern approaches to infinitesimals. In the Euclidean application, a point is an indivisible disk (the "final" disk in a sequence of disks with *vanishing* diameters). In the modern application, we obtain a disk with infinitesimal diameter (the "final" disk in a sequence of disks with *positive decreasing* diameters). Either way, BMI is involved, but it applies to different articulation of the generative sequential processes. But, I argue, these applications of BMI are far from exhaustive with respect to notions of infinitesimals and indivisibles.

Seventeenth-century mathematicians indeed debated whether the indivisible elements that lie on a surface are Euclidean lines, or rectangles with null width (which, according to the preceding BMI reconstruction, should have been identical to the former, since BMI reconstructs Euclidean points as disks with zero diameter), or rectangles with nonzero width smaller than any finite magnitude. But they also argued whether and how surfaces were generated from indivisibles. Indivisibles were alternatively conceived as fixed components whose union exhausted surfaces, as stationary elements that do not exhaust surfaces unless they are slightly extended, and as dynamic elements that exist as moving through the surface that they generate. The first and second of these may be consistent with an underlying BMI, but not the last one (associated with, say, Cavalieri; see Stedall 2008, 63–65). The notion of indivisible turns out to be underdetermined and relating to a large variety of different irreconcilable approaches (Boyer 1959, ch. 4).

This diversity is further demonstrated by the multiple non-Archimedian models (models that allow infinitesimal quantities). Lakoff and Núñez account for two of them: Robinson's nonstandard model (applying the BMI to shrinking intervals of numbers while forcing the resulting entities to obey the real numbers axioms, generating a field of hyper-real numbers that include infinitesimals) and the "granular" numbers that they prefer (applying the BMI to numbers to obtain a first order of infinitesimals, and then allowing subsequent higher orders of smaller infinitesimals, but without respecting all the axioms of the real numbers).

Lakoff and Núñez consider the latter model (which is their own ad hoc invention) more natural, but, as their measure of naturalness is BMI, rather than actual practice, they ignore its many practical problems. For example, in this model, an infinitesimal of the first order can be squared, but would not have a square root. This would make a mess of real polynomial algebra, and be fatal to the mathematical treatment of, among other things, Brownian motion.

Regardless of the respective advantages and drawbacks of the two preceding models, the plethora of non-Archimedean approaches surveyed by Ehrlich (2006) shows that the two options leave out many accounts. I'd like to highlight here the model consisting of "horn angles": angles between a circle and a tangent, as considered in Euclid's *Elements* III.16. These angles form a model of infinitesimal angles (and are indeed recognized by Euclid as smaller than any rectilinear angle) without depending on BMI for their construction.

There are also conceptions of infinitesimals that build on algebraic metaphors. For example, around the turn of the nineteenth century, French mathematician Carnot considered infinitesimals as variables:

> the quantities called Infinitely small in Mathematics are never quantities actually nothing, nor even quantities actually less than such or such determinate magnitudes; but merely quantities which the conditions of the proposed question, and the hypotheses on which the calculation is established, allow to remain variable until the operation is completed. (Carnot 1832, 19)

These variable quantities are involved in "imperfect equations," namely, equations involving a variable that require another infinitesimal (an-

other variable) in order to balance. For example, equating a small secant to a tangent of a given curve at a given point is only possible if
one allows the addition of a variable quantity that would account for
the gap between the secant and the tangent. These imperfect equations
allow mathematicians to replace curve segments by line segments and
surface elements by rectilinear shapes in calculations. Carnot argues
that if one takes a legitimately derived imperfect equation, and drops
the variable part, one gets a correct equation in the standard sense.

But here is the catch: Carnot's metaphor is not that of discarding
small or evanescent elements, but that of discarding the imaginary part
from a complex equation (that is, if $a + bi = c + di$, then one may legitimately discard the imaginary parts to obtain $a = c$):

> If I ask you what is the meaning of an equation in which imaginary quan-
> tities enter ... you answer, that this equation can only assist in discover-
> ing the true value of the unknown quantity, when by any transformations
> whatever we have effected the elimination of the imaginary quantities. I
> make the same reply for my valueless quantities [variable infinitesimals]:
> I employ them only as auxiliaries: I allow that my calculation is not rigor-
> ously exact, until I have eliminated all: until that time it is not complete,
> and does not admit of application. (Carnot 1832, 35)

Carnot's infinitesimals, constructed via an algebraic metaphor that
has nothing to do with BMI, involve "valueless" quantities rather than
evanescent ones.

Another algebraic approach is presented by Euler. For Euler, infinitesimals were algebraic zeros (Euler 2000 [1755], 51). As a result, a finite quantity plus an infinitesimal was equal to the original finite
quantity. But since the ratio between zeros equals any other finite ratio
(according to the rule of cross multiplication—for example, $0:0 = 1:2$
because $0 \cdot 2 = 1 \cdot 0$), different zeros can have different ratios to each other.

Here the trick is to go from thinking about zero as a single entity to
thinking about it as a genus of multiple species. Euler's lead allows
us, in turn, to think of infinity not only in terms of BMI, but also as an
arithmetic $1/0$. If we have a notion of "relative nothing," it is easy to
construct a formally symmetric notion of "relative everything," namely,
an algebraic notion of infinity that allows to consider different orders
of infinity.

Thinking of orders of infinity, there is yet another notion of infinity that is not reducible to BMI: uncountable, strongly inaccessible cardinals. To explain this term somewhat loosely, suppose you have a notion of an infinite cardinality (say, the least infinite cardinal $\aleph_0$). Consider all subsequent orders of infinity that can be constructed using any operations with this already existing order of infinity. For example, taking a power set allows us to construct from $\aleph_0$ the cardinality $\aleph$, the continuum. Strongly inaccessible cardinals are defined in such a way that they cannot be constructed from smaller cardinals, namely, they cannot be approached by any infinite process constructable from $\aleph_0$. In a way, they are conceived, precisely, as that which cannot be conceived through BMI! Note that every inaccessible cardinal is also a limit cardinal—I will not define this term here, only note that it is amenable to some BMI account. However, inaccessible cardinals are conceived and defined not as entities that are the final state of some indefinite process, but as entities that are beyond any final state of any indefinite process. So BMI can be projected on inaccessible cardinal post-hoc, but this has nothing to do with their cognitive origin.

But BMI is not only underdetermined and insufficient, it also fails to live up to what Lakoff and Núñez consider as the definitive feature of mathematical metaphors: it fails to obtain best fit with the relevant inferential structure. Indeed, BMI is supposed to carry as many inferences as possible from iterative finite processes to indefinite ones (for example, the existence of a final state). But iterative finite processes not only have a unique final state that is to be imposed on indefinite processes; iterative finite processes also have a unique penultimate state—something that most concepts of infinity do not have. How come the penultimate state is not carried over by the metaphor?

Indeed, it could be. Our basic notion of infinity could have been the ordered structure $1,2,3, \ldots ,n, \ldots ,n', \ldots ,3',2',1'$, where the primed integers are ordered inversely to nonprimed ones, and all primed integers come after all nonprimed ones. Indeed, there we'd have a final state $1'$, a penultimate state $2'$, and so on, improving the fit with finite intuitions. Nevertheless, this model, despite its improved fit, is not very popular (it can be historically related to Fontenelle's 1727 highly criticized attempts and to the contemporary computational model of Sergeyev 2013, which has not (yet?) gained wide support). This shows that maximal preservation of inferential structure cannot be an adequate description of what mathematical metaphors do in general, or BMI in particular. Mathematical metaphors are subject to other constraints that support the endorsement of other structures (such as simplicity).

So we see that BMI does not cover all notions of infinity and is selective in the way it transfers inferential structure. To make one final point, let's consider yet another example. Lakoff and Núñez explain the projective points at infinity (the ideal intersections of parallel lines in projective geometry) by means of the increasingly receding intersection point of almost parallel lines, in line with BMI. Henderson (2002) observes that this explanation should yield two intersection points at infinity for any pair of parallel lines—one point in each direction.

So BMI cannot be the origin of the single projective point at infinity. Indeed, the origin of this point at infinity is easy to track down: it comes from the vanishing point in Renaissance art and draftsmanship—a practical technique, rather than an inference carried between mathematical domains. Here, the relevant constraint is the alliance of mathematics with other vocations.

There are no monopolistic explanations of all notions of infinity and limit. We might be able to rearrange many concepts in terms of our preferred Ur-concept, but doing that would be a normative project (and a problematic one at that) rather than a descriptive tool of mathematical cognition theory. This applies to most mathematical concepts. Mathematical concepts survive and evolve due to the interaction of various metaphors, practices, signs, and tools subjected to an intricate system of constraints. It cannot be captured by monopolistic reductions.

The fact that formal logical presentations of mathematical knowledge can reduce concepts to a rigorous linear chain going from foundations to higher constructs does not imply the alleged conclusion that mathematical thought depends on such a linear chain. In fact, mathematical thought depends on a not always coherent superposition of practices and constraints. The challenge of the modern mathematician is to combine the different dimensions and relations of her or his ideas into piecemeal formaliz*able* definitions and proofs. But these proofs and definitions are an end result, not a description of mathematical cognition.

CHAPTER 7

Making a World, Mathematically

To conclude this book, I want to bring up an aspect of mathematical practice that was not salient enough in the narratives that I've put together so far. So far, I have been describing mathematics as responding to constraints (natural, social, practical, cognitive ...). What's missing is how mathematics changes the reality in which it evolves, feeding back into the constraints that shape it.

To do that, I am going to follow a path that might cause some raising of eyebrows: I will follow three nineteenth-century post-Kantian German thinkers: Fichte, Schelling, and Hermann Cohen. This may seem odd, because their German idealism and neo-Kantianism are usually seen today as obscure oddities, which may be relevant for historians of philosophy, but not for contemporary philosophizing as such.

There are some good reasons for this evaluation. Nineteenth-century German philosophy is at odds with some of the discursive and logical standards of contemporary mainstream philosophy, and its cultural-intellectual context is so foreign, that readers have to immerse themselves in reading for quite a while before it starts making sense. The problem is further exacerbated when idealism is presented as claiming that reality is nothing but a product of our minds and that empirical experience is therefore an illusion (reflected by the famous anecdote about Samuel Johnson's "refutation" of Berkeley's idealism by kicking a stone). This straw-man account reduces idealism to arbitrary conceptual accounts imposed on reality.

But the task of post-Kantians thinkers was much more challenging. Since they took empirical science very seriously, they sought to reconstruct the kind of conceptual account that they considered *necessary* to explain how human thinking could match empirical experience, while still retaining the relative autonomy of reason. They were after the conceptual structure without which, they believed, our experience would fall apart into incoherent strata of sensation and thought. Given my apprehension that contemporary philosophy of mathematics is stuck in a rut of realism-nominalism debates that are no longer very productive, it might be useful to pick up on those abandoned threads of thought and see where they might lead. The revived interest of analytic philosophers in Kant and Hegel (via the Pittsburgh school) indicates that my attempt here is not entirely out of touch with contemporary strands.

When I suggest to pick up those threads of thought, I don't do so because I believe that they provide answers or clean up our concepts (frankly, I don't think that answers or clean concepts are philosophy's most important products). I suggest picking up those threads, because I see philosophy as a discourse that's meant to nudge our thinking out of stagnation. Philosophy needn't necessarily be about telling the truth; rather, it should be about suggesting new roads of thought, where current ones have been worn out and keep leading us in circles or into dead ends. Picking up some old abandoned strands of thought is not about reviving or vindicating them. It is about appropriating them to provide inspiration for new and relevant ways of thinking.

In my particular case, looking into nineteenth-century post-Kantian thought led me to rethink the problem of the seemingly miraculous success of mathematics in describing natural phenomena (Wigner 1960), and to suggest a response to Mark Steiner's (1998) articulation of this problem: how is it that analogies based on intra-mathematical motivations end up producing mathematical structures, which fit physical phenomena that had nothing to do with these structures' original context of discovery? The approach I suggest does not follow or apply nineteenth-century post-Kantianism as such; still, it is definitely inspired by it. But let us set this aside until the last section of this chapter.

Since my purpose is not to reconstruct historically the ideas of the thinkers that I'm going to review, I allowed myself to offer very brief narratives based on specific relevant interpretations, rather than be faithful to the thinkers' own context and motivations. In the case of

Fichte, I follow Wood (2012) and Breazeale (n.d.). For Schelling, I complement Breazeale's account, which focuses on Schelling's philosophy of identity, by Gare (2011), which focuses on his philosophy of nature. As for Hermann Cohen, I follow his *Das Prinzip der Infinitesimal-Methode* (1883). Marc de Launay's introduction to his French translation of Cohen's work (1999) and Scott Edgar's (2014) paper provide useful interpretations.

Fichte

Our starting point for this narrative is Kant's understanding of geometric practice. A simplified narrative of this practice is as follows. First, one constructs a geometric form according to certain rules in pure intuition. This means that whatever is constructed is presented not on paper or in symbols, but inside our nonempirical, pure form of space—that is, the structure of experience that we impose on empirical spatial observation. This pure intuition, as Kant understands it, constructs objects out of strictly spatial magnitudes and relations. Once we have constructed such objects, we can study the features that necessarily follow the construction. The main point is that this practice is not just about dealing with concepts; it is about active construction and a way of seeing.

For Kant, the structure of a priori spatial intuition, derived as a solution to apparent antinomies of reason, is a necessary requirement for our ability to reason scientifically. If we accept Kant's analysis, and acknowledge a universal capacity to reason scientifically, this a priori structure turns out to be not only necessary, but universally true. This, in turn, allows us to contemplate universal mathematical truths in the particular constructions contained in pure spatial intuition.

Fichte's first objection was that this a priori spatial intuition is not enough to ground geometry. If all we have is this a priori intuition of a pure realm of magnitudes, everything we position there is relative and underdetermined. Indeed, pure intuition lacks grounding in location (everything that we construct can be displaced in pure intuition without consequence) and in directionality (horizontality, verticality, and hence, according to Fichte, also a clear notion of right angle). In order to provide a proper grounding to our construction, we need to impose on this intuition a frame of local and directional reference (Wood 2012, 90).

According to Fichte, this frame of reference consists of a fundamental infinite horizontal line and a fundamental infinite vertical line that intersect at a fundamental point of reference (like the axes of a Cartesian plane). The fundamental horizontal line is the diameter of an infinite circle that circumscribes the entire space (Wood 2012, 110). This frame is constructed by the thinking subject. Since, according to Kant, a priori spatial intuition cannot contain something infinite (it can only construct indefinite increase), the required fundamental construction cannot be obtained by pure spatial intuition. It therefore has to be constructed in a more fundamental sort of intuition—a nonsensual, so-called intellectual intuition.

This "intellectual intuition," abstracted not only from any empirical experience, but also from any pure sensory structure, diverges from Kant's doctrine. It adds to the Kantian architecture an intuition where reason may *construct* its concepts (as we construct shapes in the pure intuition of space), rather than just contemplate their logical possibility. To understand this difference, recall (from the second section in chapter 1) the Kantian distinction between a conceptual triangle, which is only subject to analytic logical derivations that follow from definitions, and a constructed triangle in the pure intuition of space, where one can construct further auxiliary lines that allow proving new necessary synthetic truths, such as the one about the sum of angles.

Fichte considers the concepts that he constructs in the intellectual intuition as necessary concepts. Indeed, if we fail to construct the fundamental lines, then, according to him, we will have no frame in which to think mathematically about the world. These constructions are therefore a necessary condition for our ability to apply mathematics to science. But Fichte's intellectual intuition is not just about mathematical entities. It is inspired by mathematical constructions, but includes much more general ontological constructions, including the "I" and the intersubjective universe that contains it.

According to Fichte, each philosopher must realize his concepts in construction. But since these constructions are performed in an intuition that abstracts from every contingency of experience, Fichte concludes that they are universally necessary for anyone who chooses to philosophize. The model here, again, is mathematical truths, which necessarily follow for anyone who makes mathematical constructions in Kant's pure spatial intuition, which is abstracted from any individual contingency. Indeed, Fichte's *Wissenschafstlehre* recommends the study of geometry as training for the study of philosophy, and he was sometimes referred to as "the Euclid of philosophy."

We see here how mathematics (or, rather, a philosophical image of mathematics) shapes a more general philosophical system. Mathematics here is not merely reactive. It is productive in two ways: as an inspiration for a philosophical system, and inside this system, as a universally necessary creative activity.

There is, however, an important difference between mathematical constructions and Fichte's more general ontological constructions. While geometry starts with the construction of a form and then derives mathematical products, the philosopher

> starts with the "products" (space, time, the causal structure of the sensible world, for example) … and then attempts to execute his construction by starting with the only material available following the initial act of global abstraction: the field of pure self-directed self-activity. (Breazeale n.d., 12)

What one constructs in intellectual intuition are the concepts necessary to explain the given "products" (space, time, and so on). These concepts, in turn, are shown to presuppose other concepts that need to be constructed to make sense of the whole.

It is crucial for our argument that Fichte considers his constructions not only as necessary, but also as real. This reality, however, is not necessarily about correspondence to external objects. Fichte does demand that the final products of philosophical constructions (for example, the I, the law-governed world) be measured by their correspondence to lived experience; if they fail to correspond, then philosophy failed to provide what it sought: the concepts necessary for rational engagement with experience. However, "the same is not true of the *elements* employed in this construction— the various, discreetly posited acts of the I and the various preliminary products of the same.… There is nothing outside of our construction that corresponds to these elements and to this process" (Breazeale n.d., 32). The correspondence is sought at a global level, not the level of detail.

To understand this claim in a mathematical context, one can think of Bombelli's solution to cubic equations (chapter 2). The cubic equation and its solution are real, and have counterparts in experience. But the construction of the solution goes through roots of negative numbers—conceptual entities that have no counterparts in Bombelli's geometric experience of the problem. Similarly, the symbol x in the solution of combinatorial problems by generating functions (chapter 4, discussion around figure 6.4) adds a mathematical element that does not correspond to anything in the experiential counterpart of the combinatorial problems.

So where is the reality of Fichte's constructions? The constructed concept of the I and the act of construction are, according to Fichte's *Attempt at a New Presentation of the Wissenschaftslehre*, "one and the same, and when we think of this concept, we do not and cannot think of anything but what has just been described. It *is* so, because I *make* it so" (quoted in Breazeale n.d., 34). Since construction of concepts in intellectual intuition is an actual act of an actual subject, and since the constructions subsist in this act of construction, the constructions are real.

The elements of Fichte's construction are the free creation of the philosophizing subject. They are necessary to make sense of the world, and are real as actual actions of an actual subject, but are not given in advance and do not necessarily have empirical references. Still, due to the matching between experience and intellectual construction at the level of overall result, we may think of empirical experience *as if* it were structured by the philosophical construction. This, again, parallels the practice of the geometer, who

> by virtue of the self-evidence of his demonstrations, is entitled to view the various actual figures he encounters within experience "as if" they had been constructed in the same way as were those pure figures constructed by him in imagination (even though he knows very well that they were *not* constructed in this way) and to assume that whatever is found to be necessary in the case of the latter will also be necessarily true of the former. (Breazeale n.d., 33)

We end up with a series of real acts of intellectual constructions, which match overall experience, but are free to follow their own internal logic of detail. Even though they match empirical experience only at the global level, they can still be applied to understand reality.

I won't try to solve the tension between the freedom and necessity of constructions in Fichte's thought, or explain how a construction in intellectual intuition is different from reasoning with hypothetical concepts, or discuss the viability of abstracting from everything that Fichte claims to abstract from. Rescuing Fichte from these difficulties is not my purpose here. Instead, what interests me here is the relation between an ideal reality—the real autonomous actions of a thinking subject—and given empirical experience. In Fichte's system, the former must, at some level, fit the latter, but is not isomorphic to it. It partakes in a different kind of reality, subject to a different set of constraints. It shapes a conceptual world that is necessary to make sense of the empirical one.

If, instead of saving Fichte, we wish to bring him to bear on the philosophy of mathematical practice, we can liken his free constructions to the real activity of the mathematician. Mathematical constructs are real not because they are all descriptively adequate, but because they are identified with the act of construction—the real mathematical activity of doing things with mathematical signs. If we understand mathematical reality as this reality of practice, we part with many of Fichte's abstractions and claims to universality, but we remain with a very real notion of mathematical practice.

This will not solve Steiner's challenge, because Fichte starts with empirical phenomena, and retraces the necessary concepts, whereas Steiner notes that we sometimes follow internal mathematical motivations and stumble upon models of empirical reality. But Fichte's outline does make sense of the experience of creativity *and* necessity that mathematicians testify to. It is real because we really do it this way under some very real constraints, and it is creative because it is we who come up with these activities (the kind of creative reality manifested, for example, by formal variables in generating functions and by complex numbers in the solution of real equations). This is the kind of reality that is experienced in actually solving mathematical exercises, as opposed to the sense of a made up artificial language sometimes experienced by observers who are not engaged in doing mathematics. The reality of application supports mathematical constructions only at the edges—their interaction with empirical reality—not in the core of their actual practice.

To conclude this discussion of Fichte, I want to make one crucial note. Not every experience can be reconstructed in Fichte's intellectual intuition. Some contingent and relative experiences or feelings remain inaccessible to philosophical construction, which begins by abstracting from the contingencies of experience and the specificity of a single subject. According to Fichte, this is the original limitation of the I. This limitation is not some mere dispensable background noise. This limitation defines the I as finite (Breazeale n.d., 34). The ideal construction of the philosopher is limited in its capacity to match empirical experience, and cannot construct itself without acknowledging this limitation. For Fichte, as for the mathematician, the reality of the self-constructed I is no substitute for the reality of given experiences.

Schelling

While Fichte posited two realities (one experienced as given and one self-constructed) that must fit "at the edges," but otherwise follow different paths, Schelling was much more ambitious in his attempt to match thought and being. In order to reconcile the freedom of the subject with the givenness of nature, he required the philosopher to abstract not only from experience, but also from the confinement of thought to an intellectually intuiting subject. Philosophizing would then no longer be the act of an I, but of a "pure subject-object" (Breazeale n.d., 50).

The very possibility of this act assumes an integration of objective and subjective realities as aspects of a unified whole. This is not an assumption that can be proved, according to Schelling, but a starting point for any system that hopes to affirm a unified world, rather than two worlds sewn together at the edges—one subjective and one objective. Without this assumption, according to Schelling, our science is doomed to fail, as it is not properly grounded (Breazeale n.d., 59).

Assuming this integration, there are two trajectories for philosophizing. One is "natural philosophy," which begins from the objective and derives the subjective. The other is "transcendental idealism," which follows the reverse path. The hypothesized integration of nature and thought is supposed to guarantee the correspondence of these two trajectories.

More specifically, the purpose of transcendental idealism would be "to explain the idea of an objective world which was absolutely independent of our freedom, indeed which limits this freedom, by a *process* in which the I sees itself as unintentionally but necessarily engaged, precisely through the act of self-positing" (quoted from *The History of Modern Philosophy*, Gare 2011, 43). Natural philosophy, in turn, would explain the idea of a subjective world by starting with a self constructing nature encountering internal conflicts and forming internal divisions. Idealism and realism would be complementary narratives in a unified ideal-realism (Gare 2011, 58–59).

This structure required another break from Fichte. Fichte assumed a world of given phenomena and a constructing subject. For Schelling, however, both the subject and nature were constructive. The common principle of the subject and of nature is that of "self activity positing itself and coming to be through limiting itself and being limited" (Gare 2011, 47). How does a subject/object division emerge in such a system? Schelling begins his philosophy of nature with a dynamic natural process, and derives a "multiplicity of actants [that] reciprocally restrict themselves (prehending each other) to effect the unity of a product" (Gare 2011, 49; one can think here of a cloud of gas in space evolving into a system of planets orbiting a star). The subject/object is a division derived from an integrated dynamic process in order to make sense of the whole, not a given starting point.

Since the conscious subject is derived from within nature, its own constructions form part of the construction of nature, rather than simply reflect it. Therefore, according to Schelling's *First Outline*, "to philosophize about nature means to create nature" (Breazeale n.d., 48). Indeed, we intervene in nature based on our conceptual judgments (Gare 2011, 45), so intellectual constructions have natural implications. This formulation can also be reversed: our embeddedness in nature impacts our intellectual constructions, so the constructions of nature have conceptual implications. They are all part of the same integrative process.

I'm well aware that this may sound mystical to the modern ear. Many readers considered it esoteric and circular in Schelling's lifetime as well. This effect is further aggravated when Schelling goes from his philosophy of nature to his philosophy of identity. While the former was fallible, dependent on empirical observation and limited in reach, the latter simply discarded as unreal any contingent phenomena that couldn't be subsumed into the philosophical system (Breazeale n.d., 68), bringing Schelling dangerously close to what I presented earlier as the straw-man of idealism.

My point in bringing Schelling into this narrative is to suggest that conceptual (and mathematical) thinking is part of the world that it purports to describe, and that it intervenes in this world through its products (tools, practice, technology). They shape our behavior in the world, and, as a result, transform this world. Mathematical reasoning and the world in which it takes place are co-constructive.

We can think here of the Black-Scholes vignette from the introduction: if we accept that the partial correspondence of the prices of options and the Black-Scholes formula is, at least in part, a self fulfilling prophecy, then mathematics clearly shapes the world we live in. Since the products of math dependent science and technology seem to change the earth's atmosphere, this effect is not confined to social reality. Mathematical practice, then, is not only a real activity, as in Fichte, but also has real impact on the world we inhabit. (If we think, along with Maddy, about fundamental mathematical truths as those hypotheses that turn out to be the most mathematically productive, we gain another sense of how new concepts can shape new realities.)

In Fichte's worldview, mathematical concepts fit the world because they are designed in advance to fit at some integral level, leaving the internal organization to the autonomy of human reasoning. In Schelling's view, mathematical and natural constructions conform because the phenomenal world and mathematical practice are two elements articulated from a single dynamic system: the system of human existence in the world. Within this dynamic system, they constrain each other: the world constrains mathematical practice and thought, and mathematics produces artifacts, practices and perceptions that reform the world. The relative autonomy of subject and object is derived from the self organization of a unified natural-rational process into relatively articulated components.

If, as earlier, we neglect the necessary and universal aspect of Schelling's abstractions, we end up with an image of mathematics that is much in line with the constraints based approach to the philosophy of mathematical practice suggested in chapter 3. Mathematics emerges from negotiating a continuum of subjective and objective constraints in a dynamic life world. From within these constraints, it emerges as a semi-autonomous system of knowledge. But it also impacts the world in a way that reforms and rearranges these constraints in a manner that can produce new systems of mathematical knowledge.

Hermann Cohen

Hermann Cohen provides us with a related account in a more specifically mathematical-scientific context. His approach is similar to Fichte's, but the details are somewhat different, as he is concerned with the unity and continuity of phenomena.

Cohen believes that the pure schemes of space and time do not provide an absolute unit (we can use different scales of measurement, break a unit into subelements and group multiple elements to form new units), do not allow us to conceive of curves as unified entities (as opposed to arbitrary collections of segments or points), and do not account for the notion of continuity (how points are connected to each other in a continuum, how infinite processes reach their limit). Kantian spatial and temporal intuition, according to Cohen, only provides us with relative unities and discrete collections (Cohen 1883, §31, 37, 58).

In order to adhere to the Kantian system, Cohen seeks the solution in Kant's notion of "intensive magnitude." Kant seems to use this term in two senses. The first sense opposes intensive magnitudes to extensive magnitudes. The latter, like length, can be divided into units and added by aggregation. The former, like temperature, cannot be broken up into units and aggregated. For instance, you can't break water at 50 degrees Celsius into 50 units of water at a temperature of 1 degree each, and if you pool together two buckets of water at 50 degrees, the water temperature will remain 50 degrees.

The second sense of intensive magnitude is rather different, and has to do with a degree of reality—how conceptually articulated a given experience is. To make sense of this we must note that Cohen's notion of reality (*Realität*) is not the notion of empirical givenness (*Wirklichkeit*). Reality is derived from the Latin *res*, meaning "thing." It refers to how conceptually articulated something is, distinguishing a thing from a vague experience. On the one hand, we can experience something as given without being able to understand it as clear and distinct object (such as unarticulated background noise); on the other hand, something constructed by the creative imagination (such as a winged horse) can be perfectly thing-like, or real, without being empirically given. Reality, in this specialized sense, must have something to do with thought.

The following paragraph illustrates this notion of reality:

> Nature is not something given there in itself, which requires an application of our mathematical description; rather, it is first discovered and produced by these descriptions. If conic sections [for example, ellipses] had not been thought of, we would not have known the natural process formed by the orbits of planets. In that case, this process would be at best a problem, perhaps the problem of epicycles [referring to the classical attempt to model planet motion by superposing circular motions]. This product of pure geometry [the conic sections] is not applied to the nature of planet orbits given in itself, but is used to produce this nature, which, without such pure means of production, would be given only as a question, not as certified natural process. (Cohen 1883, §91)

Planet orbits only become scientific entities once they are articulated rationally. Before science manages this articulation, planet orbits are but an imperfectly conceived challenge: discrete glimmers observed here and there, never given as unified continuous wholes. They can be roughly related to the notion of epicycles, but not properly grasped by it. Planet observations are empirically given, but without conic sections they are not rationally formed as clear and distinct orbits, and are therefore not endowed with reality in Cohen's sense.

When Cohen considers intensive magnitudes, he prefers to understand them as measures of the reality of phenomena rather than as intensities of sensation (heat, light, and so on). This is not an unproblematic interpretation of the Kantian text, but one that Cohen is firmly committed to: "The degree [the intensive magnitude] is a determination not so much of sensation as of pure thought" (Cohen 1883, §78). The intensive magnitude of something given empirically establishes it as an articulated thing, that is, as real. But this intensive magnitude itself is a product of reason, not something given.

So what exactly is this mysterious intensive magnitude that makes scientific experience real? According to Cohen, it is the infinitesimal. Indeed, the infinitesimal is not given in experience. Rather, it is a product of reason that articulates natural phenomena as things, therefore endowing them with reality.

How does the infinitesimal do that? First, according to Cohen, the infinitesimal is not a relative quantity attributed contingently to extensive magnitudes (like articulations of magnitudes into units), but applies as a nonempirical measure of change. Since it does not depend on the contingencies of subjective empirical perception, it is, in Cohen's terms, absolute. The infinitesimal connects each point to its neighbors, thus forming continuity. It also connects processes to their limits. Indeed, the infinitesimal is the *difference* between a function and its limit, but since it is not present empirically, this difference can be seen as an *identification* of the end of the process with its limit (there is a strong affinity here to Wronski's approach to infinity [Wagner 2014], but I can't establish a historical connection; Solomon Maimon [see Freudenthal 2013] would have been a more accessible influence, but Cohen claimed that he only read Maimon's work later in his career).

Perhaps more importantly, the infinitesimal allows us to express curves and other physical phenomena by unified formulas. Indeed, some curves, such as circles and conics can be described by equations without resorting to infinitesimals, but differential equations can describe a much richer set of phenomena by finitely expressible equations. In fact, in eighteenth-century and early nineteenth-century mathematics, functions were sometimes defined as that which is expressed by a single formula.

Now this articulation could be read in terms of Fichte's and Schelling's constructions: we have here a concept—the infinitesimal—which is necessary in order to make sense of empirical experience. But Cohen brings another twist. The infinitesimal is not constructed by Fichte's abstract subject, nor by Schelling's even more abstract subject-object. The unified consciousness that constructs the infinitesimal is a cultural, discursive and intersubjective consciousness: science as discourse (Cohen 1999, 17). The infinitesimal is validated by its necessity for turning empirical experience into a reality grounded in communal rational discourse, rather than in an abstract rational intellectual intuition.

Cohen's praise for the infinitesimal is somewhat reactionary with respect to his contemporary mathematics. The late nineteenth century is the time of the arithmetization of geometry, the discretization of the real line into a set of points, and the rejection of infinitesimals in favor of epsilon-delta formulations. It may be that Cohen is worried that his contemporary science is losing something crucial as it discards the infinitesimal in favor of an arithmetization that allows all sorts of non smooth "monster" functions to roam free.

Cohen's anachronism explains, perhaps, why Ernst Cassirer, Cohen's student and colleague, preferred relations and functions over infinitesimals as his realizing concepts for science. But either way, the neo-Kantian reality of these concepts is not just about given experience, it is about given experience articulated into real things by discursive concepts grounded in thought rather than sensation. Without such concepts, our sensory experience would be the experience of unarticulated noise (discrete glimmers, rather than orbits modeled after conic sections). Concepts therefore articulate reality into unities and continua, rather than just reflect it as is.

The controversy between Cohen and Cassirer brings about the issue of relativity: is there a unique necessary conceptual framework for reality, or can we have various competing frameworks? Cohen's position seems to tend toward necessity and uniqueness, at least as a regulative idea. Cassirer's later work on symbolic forms acknowledges that several alternative frameworks can realize our experience (myth, religion, language, science). There is a clear sense of advance as Cassirer compares these systems historically, but also an apparent recognition that they cannot refute each other.

If we want to push further in the direction of relativity, we can turn to Harold Garfinkel's ethnomethodology as applied to mathematics by Eric Livingston. The ethnomethodological approach emphasizes how people make sense of social situations without assuming that all social rules are given in advance. People are obviously subject to social constraints, but often have room to make up some rules as they go along. Livingston (1986, 1999) emphasizes the contingencies of historical and modern practices of proving and shows how different historical mathematical discursive frameworks were constructed to establish mathematical claims as true. The realizing power of mathematical conceptualization, in this view, does not lead to a unique framework.

But regardless of the question of relativity, Cohen's narrative presents us with a mathematics that is not only real because it is implemented in action (Fichte) and is co-constitutive of empirical reality (Schelling), but because it turns unarticulated experience into an articulated set of phenomena with clear continuities and individuations. If we allow ourselves (*pace* Cohen) to superpose the discursive articulations of science onto individual cognitive processes, we come close to the picture presented in the fourth section of chapter 5: conceptual articulations shape perception by instigating preafference and as operators that constrain the way stimuli give rise to brain dynamics (or attractors), thus producing various conceived realities.

These three approaches point to a focus on mathematical practice and its interaction with the world, rather than on mathematics as a set of real or nominal abstractions. Despite their obscurity, these approaches have much more to do with how we learn, produce, and use mathematics than some of the most dominant expressions of contemporary ontological and logical debates in the philosophy of mathematics.

The Unreasonable(?) Applicability of Mathematics

The previous sections were not meant to teach nineteenth-century post-Kantian philosophy. They were meant to suggest that we can make sense of the idea that thought constructs reality without succumbing to the straw man of idealism. One sense is Fichte's, where a subjective reality corresponds to given experience only at an overall level, while its details follow a different kind of autonomous evolution; it is nonetheless real, because it is the product of the actual work of an actual subject, and is required for a rational understanding of given experience. Another sense is Schelling's subjective-objective reality, reminding us that thought may lead to action that changes nature, and vice versa. This view endorses a dynamic system of nature and thought that constantly rearticulate each other while organizing themselves into relatively autonomous subsystems. Finally, there's Cohen's inter-subjective reality, where discursive concepts are crucial for us to make scientific sense of things—concepts without which our experience will remain noisy and vague, not worthy, according to Cohen, of the adjective "real."

This summary should not be read as advocating idealism. It should be read as an attempt to argue that the term "reality" is too rich to be reduced to a force constraining us from outside and opposed to autonomous creations of the mind. The post-Kantian tradition is important because it offers various more complex articulations of the relations between being and thought. It argues for a reality of ideas that is continuous with the reality of nature.

This reality of ideas means that they really impact our world and help shape it. If we think of mathematics as a cultural negotiation of constraints (natural, social, practical, cognitive …), mathematical ideas feed into these constraints and take part in reshaping them. So to say that thought shapes reality is not to claim that we make everything up as we go along, or that we simply impose our thoughts on reality.

This line of thinking encourages us to focus on the mathematics as a real activity, as something that interacts with the world, and as something deeply embedded in our scientific sense of realness. It encourages engaged, practice-oriented mathematics, as opposed to viewing mathematics as a formal game, a nominal abstraction or a reflection of a fully given outside world. It imposes an ethics of mathematical engagement as opposed to a detached form of objectivity that only states truths without taking any responsibility for their formation and impact.

Now, if mathematical constructions are real in the preceding senses, then the effectiveness of mathematics in the natural sciences is perhaps not so unreasonable as claimed by Wigner. Mathematics is designed to fit, at some level, natural phenomena (Fichte); it is embedded in our technology that shapes new phenomena (Schelling); and it defines what we can take seriously as a scientific reality, rather than just unarticu-lated noise (Cohen).

In many ways, the relation between mathematics and phenomena should not be more surprising than the relation between language and phenomena: many phenomena can be represented by language (of course, not any linguistic description fits any phenomena, but this is true of mathematics as well); language is a good framework for produc-ing reasonable conjectures about the world; language helps us commu-nicate and thereby reshape the world; and whatever is not linguisti-cally describable is factored out of the discursive system of science.

But this does not quite resolve the problem of the surprisingly suc-cessful application of mathematics to empirical phenomena. Mark Steiner's (1998) take on the problem invokes several examples of math-ematical constructions, motivated by intra-mathematical analogies and anthropocentric pragmatic and aesthetic constraints, which end up providing precise models for physical realities far removed from the anthropocentric constraints that shaped them. It's not that all, or even most mathematics is empirically useful, but that mathematics as a whole, despite its human origin, is an efficient system for formulat-ing surprisingly successful scientific conjectures. Steiner considers this as evidence for the anthropomorphic design of the universe—a design of the universe in the image of human reason. According to Steiner, such design explains how the universe is amenable to description by mathematical models that evolve from the internal aesthetics of human reasoning, and that needn't have anything to do, a priori, with macro- or microphysical processes. This design might suggest some superven-ing divine or natural reason. I'll try to offer a more secular explanation.

Several scholars (Carrier 2003; Maddy 2007, 335–38; Simons 2001, 183) cast doubt on the break that Steiner postulates between his exam-ples of human mathematical constructions on the one hand and the physical context to which they are supposedly surprisingly applied only post hoc. But this is not the direction I want to take here. I would like to explain why it is, after all, reasonable, that a mathematical con-struction that evolved in one context can end up applying to other, unrelated contexts.

The answer can't be "statistical" either. There are indeed many math-ematical structures, and many phenomena, and there is some chance that they may accidentally correspond. But since we don't have a quantitative model for analyzing this probability (What constitutes a "single phenomenon" and a "single mathematical model"? How many phenomena and models are there? What is the stochastic process by which we look for a model that fits a given phenomenon?), this is a dead end. For all we know, the chance of correspondence may be smaller than the chance of a typing monkey proving a complicated theorem.

Yet another attempt can suggest that there is some inherent similarity between very different natural phenomena, which makes them likely to be captured by variations on existing mathematical models. But similarity, as argued by Steiner (1998, 53), is a very underdetermined term. What kind of similarity can we articulate between the motion of pollen and stock prices, both describable by means of Brownian motion, other than the similarity of the mathematical models that manage to describe them both? If we simply suggest post hoc that there had to be some internal underlying similarity preexisting the common mathematical modeling, we're begging the question. And even if there is some sort of underlying nonmathematical similarity between these phenomena, there's no reason why it should be captured by *mathematically* similar models.

So if we want to explain the applicability of mathematics based on a notion of similarity, we have to articulate this similarity in a more precise manner. Similarity cannot apply to phenomena as such. Instead, it should apply to the scientific articulations of phenomena. So let's follow this thread.

Indeed, dwelling in the world as scientists do is not about a passive observation of phenomena. It is about channeling phenomena via instruments of observation and measurement into mathematical forms of inscription. Our instruments of observation and measurement already embody mathematical relations—they are constructed and calibrated based on mathematical knowledge. Mathematical forms of inscription further impose mathematical relations on recorded phenomena. Observed scientific phenomena are therefore already partly realized (in Cohen's sense) or constructed (in Schelling's or Fichte's sense) by means of mathematical concepts. These realizations and constructions are no less restrictive than linguistic encoding; and just as linguistic encoding imposes a restricted array of similar grammatical relations on very different sentences, so does mathematical encoding reduce phenomena to a field of restricted mathematical relations. What's not expressible within these constraints is simply factored out of quantitative science.

In this restricted context of mathematically shaped and encoded observations (rather than "raw phenomena"), the notion of similarity between different collections of observations is more clearly defined. Assuming that mathematized observation and formulation force empirical phenomena into a restricted domain, we can argue that similarity between diverse mathematized phenomena is not unlikely, and therefore it is not terribly surprising that different mathematically shaped and encoded observations are captured by similar mathematical models, even if the development of these models depended on intra-mathematical motivations rather than the eventual empirical application. The notions of similarity used here all relate to mathematical structures, and are therefore homologous and correlated. The mathematical similarity of mathematized phenomena is intrinsically related to the mathematical similarity of the mathematical models that capture them.

To put it in terms of another metaphor, if we share the same building tools and technologies, then it is not terribly surprising that we sometimes come up with similar buildings in very different terrains and physical conditions, as long as we note that these buildings are built only in portions of the environment that are amenable to our building tools and technologies. Analogously, if the reality that is scientifically accessible to us is indeed realized or constructed in the image of our mathematical concepts and tools, then the relatively small portion of empirical reality that we manage to realize or construct is precisely that which is amenable to our restricted set of mathematical tools and technologies, and can therefore be captured by the relatively small world of mathematical models emerging from human reason. In a way, science carves precisely that portion of our natural-conceptual-practical reality that is explorable in terms of mathematical reason.

This is not a complete solution to Steiner's problem, as it assumes some sort of "compactness" in the realm of mathematically shaped and encoded empirical observations that would make similarities likely. However, we still have no a statistical framework to establish this compactness (we still have no articulation of single elements, an overall count or a stochastic process of selection). So perhaps the problem is not solved, but only reduced: from the "unreasonable applicability of mathematics to natural science," to the "unreasonable(?) applicability of mathematical models to records of empirical observations produced by mathematically designed instruments and mathematical forms of inscription." Still, even in this view, it would be an exaggeration to claim, with Nietzsche, that scientists discover that which they had themselves hid behind the bush.

■ ■ ■ ■ ■

To put it in a nutshell, this book described mathematical practice as a negotiation of various constraints by means of rearticulated, superposed, and deferred interpretations in a contemplated and experienced reality. The practical task of the humanist in this context remains, then, to provide a detailed and contextual account of how mathematical knowledge is formed as part of our strategies of inhabiting our constrained and constraining world, building on ideas, tools, manipulations of inscriptions and on the observations mediated through them.

Different thoughts and tools that depend on combining various metaphors and analogies allow us to inhabit our world differently, to know and carve different portions of reality. I like thinking, with Steiner, of known reality as anthropomorphically designed. But I'd like to think of it not simply as a given anthropomorphic fact; I'd like to think of reality, with the many scholars, practitioners, and activists who keep experimenting with our language, knowledge, and world, as a creation that we can, to a certain extent, reform, so as to make it better.

Bibliography

■ ■ ■ ■ ■ ■ ■ ■ ■ ■

Alexander, Amir. 2014. *Infinitesimal: How a Dangerous Mathematical Theory Shaped the Modern World*. New York: Scientific American / Farrar, Straus and Giroux.

Arrighi, Gino, ed. 1974. *Tractato d'abbacho: dal Codice Acq. e doni 154 (sec. XV) della Biblioteca Medicea Laurenziana di Firenze*. Pisa: Domus Galilaeana.

Asper, Markus. 2008. "The Two Cultures of Mathematics in Ancient Greece." In *The Oxford Handbook of the History of Mathematics*, edited by Eleanor Robson and Jacqueline Stedall, 107–32. Oxford: Oxford University Press.

Asratian, Armen S., Tristan M. J. Denley, and Roland Häggkvist. 1998. *Bipartite Graphs and Their Applications*. Cambridge: Cambridge University Press.

Azzouni, Jody. 1994. *Metaphysical Myths, Mathematical Practice: The Ontology and Epistemology of the Exact Sciences*. Cambridge: Cambridge University Press.

———. 2005. "Is There Still a Sense in Which Mathematics Can Have Foundations?" In *Essays on the Foundations of Mathematics and Logic*, edited by G. Sica, 9–47. Monza: Polimetrica.

Barany, Michael J., and Donald MacKenzie. 2014. "Chalk: Materials and Concepts in Mathematics Research." In *Representation in Scientific Practice Revisited*, edited by Catelijne Coopmans, Janet Vertesi, Michael Lynch, and Steve Woolgar, 107–30. Cambridge, MA: MIT Press.

Barsalou, Lawrence W. 2003. "Abstraction in Perceptual Symbol Systems." *Philosophical Transactions of the Royal Society B: Biological Sciences* 358, no. 1435 (July 29): 1177–87. doi:10.1098/rstb.2003.1319.

Barthes, Roland. 1971. "The Structuralist Activity." In *Critical Theory since Plato*, edited by Hazard Adams, 1128–30. London: Heinle and Heinle.

Bell, E. T. 1923. "Euler Algebra." *Transactions of the American Mathematical Society* 25, no. 1: 135–54. doi:10.2307/1989055.

Berkeley, George. 1734. *The Analyst, Or, A Discourse Addressed to an Infidel Mathematician: Wherein It Is Examined Whether the Object, Principles, and Inferences of the Modern Analysis Are More Distinctly Conceived, or More Evidently Deduced, than Religious Mysteries and Points of Faith*. London: J. Tonson. http://archive.org/details/theanalystoradis00berkuoft.

Black, Fischer, and Myron Scholes. 1973. "The Pricing of Options and Corporate Liabilities." *Journal of Political Economy* 81, no. 3: 637–54. doi: 10.1086/260062.

Bogart, Kenneth P., and Peter G. Doyle. 1986. "Non-Sexist Solution of the Ménage Problem." *American Mathematical Monthly* 93, no. 7: 514–18. doi:10.2307/2323022.

Bollobás, Béla. 1998. *Modern Graph Theory*. New York: Springer.

Bombelli, Rafael. 1929. *L'algebra, Libri IV e V*. Edited by Ettore Bortolotti. Bologna: Zanichelli. http://storage.lib.uchicago.edu/pres/2005/pres2005-188.pdf.

——. 1966. *L'algebra*. Edited by Ettore Bortolotti. Milan: Feltrinelli.

Bortolotti, Ettore. 1933. *I cartelli di matematica disfida e la personalità psichica et morale di Girolamo Cardano*. Imola: P. Galeati.

Bos, Henk J. M. 2012. *Redefining Geometrical Exactness: Descartes' Transformation of the Early Modern Concept of Construction*. New York: Springer.

Bostock, David. 2009. *Philosophy of Mathematics: An Introduction*. Chichester, UK / Malden, MA: Wiley-Blackwell.

Bottazzini, Umberto. 1986. *The Higher Calculus: A History of Real and Complex Analysis from Euler to Weierstrass*. New York: Springer.

Boyer, Carl B. 1959. *The History of the Calculus and Its Conceptual Development*. New York: Dover. http://archive.org/details/TheHistoryOfTheCalculusAndItsConceptual Development.

Breazeale, Daniel. n.d. "Men at Work: Philosophical Construction in Fichte and Schelling." Preprint. https://www.yumpu.com/en/document/view/19583670/1-men-at-work -philosophical-construction-in-fichte-and-schelling-.

Brigaglia, Aldo, and Ciro Ciliberto. 2004. "Remarks on the Relations between the Italian and American Schools of Algebraic Geometry in the First Decades of the 20th Century." *Historia Mathematica* 31, no. 3: 310–19. doi:10.1016/j.hm.2003.09.003.

Bueti, Domenica, and Vincent Walsh. 2009. "The Parietal Cortex and the Representation of Time, Space, Number and Other Magnitudes." *Philosophical Transactions of the Royal Society of London B: Biological Sciences* 364, no. 1525: 1831–40. doi:10.1098/ rstb.2009.0028.

Burgess, John P., and Gideon A. Rosen. 1997. *A Subject with No Object: Strategies for Nominalistic Interpretation of Mathematics*. Oxford: Oxford University Press.

Byers, William. 2007. *How Mathematicians Think: Using Ambiguity, Contradiction, and Paradox to Create Mathematics*. Princeton, NJ: Princeton University Press.

Carnot, Lazare. 1813. *Réflexions sur la métaphysique du calcul infinitésimal*. Paris: Mme. Ve. Courcier. https://archive.org/details/bub_gb_ATXTgQdZjZcC.

——. 1832. *Reflexions on the Metaphysical Principles of the Infinitesimal Analysis*. Oxford: J. H. Parker. https://archive.org/details/reflexionsonmet00browgoog.

Carrier, Richard. 2003. "Fundamental Flaws in Mark Steiner's Challenge to Naturalism in the Applicability of Mathematics as a Philosophical Problem." *The Secular Web*. http://www.infidels.org/library/modern/richard_carrier/steiner.html.

Cartwright, Nancy. 1983. *How the Laws of Physics Lie*. Oxford: Oxford University Press.

Cassirer, Ernst. 1910. *Substanzbegriff und Funktionsbegriff: Untersuchungen über die Grundfragen der Erkenntniskritik*. Berlin: B. Cassirer. https://archive.org/details /substanzbegriffu00cassuoft.

Cavaillès, Jean. 1994. *Œuvres complètes de philosophie des sciences*. Edited by Georges Canguilhem. Paris: Hermann.

Cellucci, Carlo. 2007. *La filosofia della matematica del Novecento*. Rome: Laterza.

Cifoletti, Giovanna Cleonice. 1992. "Mathematics and Rhetoric: Jacques Peletier, Guillaume Gosselin, and the Making of the French Algebraic Tradition." Ph.D. Dissertation, Princeton University.

——. 1995. "La Question de l'Algèbre. Mathématiques et Rhétorique des Hommes de Droit dans la France du XVIe Siècle." *Annales. Histoire, Sciences Sociales* 50, no. 6: 1385–1416. doi:10.3406/ahess.1995.279438.

Clagett, Marshall. 1959. *The Science of Mechanics in the Middle Ages.* Madison: University of Wisconsin Press.

Cohen, Hermann. 1883. *Das Prinzip der Infinitesimal-Methode und seine Geschichte.* Berlin: Dümmler. http://archive.org/details/dasprincipderin01cohegoog.

——. 1999. *Le Principe de la Méthode Infinitésimale et Son Histoire.* Paris: Vrin.

Cohen Kadosh, Roi, and Vincent Walsh. 2009. "Numerical Representation in the Parietal Lobes: Abstract or Not Abstract?" *Behavioral and Brain Sciences* 32, nos. 3–4: 313–73. doi:10.1017/S0140525X09990938.

Cohn, Carol. 1987. "Slick 'Ems, Glick 'Ems, Christmas Trees, and Cookie Cutters: Nuclear Language and How We Learned to Pat the Bomb." *Bulletin of the Atomic Scientists* 43, no. 5: 17–24. doi:10.1080/00963402.1987.11459533.

Colyvan, Mark. 2011. "Applying Inconsistent Mathematics." In *The Best Writing on Mathematics 2010,* edited by Mircea Pitici, 346–57. Princeton, NJ: Princeton University Press.

Coopmans, Catelijne, Janet Vertesi, Michaeland Lynch, and Steve Woolgar, eds. 2014. *Representation in Scientific Practice Revisited.* Cambridge, MA: MIT Press.

Corry, Leo. 2013. "Geometry and Arithmetic in the Medieval Traditions of Euclid's Elements: A View from Book II." *Archive for History of Exact Sciences* 67, no. 6: 637–705. doi:10.1007/s00407-013-0121-5.

d'Alembert, Jean le Rond. 1995. *Preliminary Discourse to the Encyclopedia of Diderot.* Chicago: Chicago University Press.

Dardi, Maestro. 2001. *Aliabraa argibra dal manoscrito I.VII.17 della Biblioteca Comunale di Siena.* Edited by Raffaella Franci. Siena: Università degli studi di Siena.

David, Paul A. 2004. "Understanding the Emergence of 'Open Science' Institutions: Functionalist Economics in Historical Context." *Industrial and Corporate Change* 13, no. 4: 571–89. doi:10.1093/icc/dth023.

——. 2008. "The Historical Origins of 'Open Science': An Essay on Patronage, Reputation and Common Agency Contracting in the Scientific Revolution." *Capitalism and Society* 3, no. 2. doi:10.2202/1932-0213.1040.

De Cruz, Helen. 2008. "An Extended Mind Perspective on Natural Number Representation." *Philosophical Psychology* 21, no. 4: 475–90. doi:10.1080/09515080802285289.

De Cruz, Helen, and Johan De Smedt. 2010. "The Innateness Hypothesis and Mathematical Concepts." *Topoi* 29, no. 1: 3–13. doi:10.1007/s11245-009-9061-8.

Dehaene, Stanislas. 2009. "Origins of Mathematical Intuitions: The Case of Arithmetic." *Annals of the New York Academy of Sciences* 1156: 232–59. doi:10.1111/j.1749-6632.2009.04469.x.

——. 2011. *The Number Sense: How the Mind Creates Mathematics, Revised and Updated Edition.* New York: Oxford University Press.

Dehaene, Stanislas, Ghislaine Dehaene-Lambertz, and Laurent Cohen. 1998. "Abstract Representations of Numbers in the Animal and Human Brain." *Trends in Neurosciences* 21, no. 8: 355–61. doi:10.1016/S0166-2236(98)01263-6.

Dehaene, Stanislas, Manuela Piazza, Philippe Pinel, and Laurent Cohen. 2003. "Three Parietal Circuits for Number Processing." *Cognitive Neuropsychology* 20, no. 3: 487–506. doi:10.1080/02643290244000239.

Delany, Sheila. 1990. *Medieval Literary Politics: Shapes of Ideology*. Manchester, UK: Manchester University Press.

Deleuze, Gilles. 2003. *Francis Bacon: The Logic of Sensation*. Minneapolis: University of Minnesota Press.

de Libera, Alain. 1996. *La Querelle des Universaux: de Platon à la Fin du Moyen Age*. Paris: Seuil.

della Francesca, Piero. 1970. *Trattato d'Abaco. Dal Codice Ashburnhamiano 280 (359*–291*) della Biblioteca Medicea Laurenziana di Firenze*. Edited by Gino Arrighi. Pisa: Domus Galilaeana.

Demidov, S. S., and A. Shenitzer. 2000. "Two Letters by N. N. Luzin to M. Ya. Vygodskii." *American Mathematical Monthly* 107, no. 1: 64–82. doi:10.2307/2589382.

Derman, Emanuel, and Nassim Nicholas Taleb. 2005. "The Illusions of Dynamic Replication." *Quantitative Finance* 5, no. 4 (August 1): 323–26. doi:10.1080/14697680500305105.

Derrida, Jacques. 1978. *Writing and Difference*. Chicago: University of Chicago Press.

———. 1988. *Limited Inc*. Evanston, IL: Northwestern University Press.

Descartes, René. 1954. *The Geometry*. New York: Dover Publications, 1954. http://archive.org/details/TheGeometry.

Diamond, Cora, ed. 1976. *Wittgenstein's Lectures on the Foundations of Mathematics, Cambridge, 1939: From the Notes of R. G. Bosanquet, Norman Malcolm, Rush Rhees, and Yorick Smythies*. Ithaca, NY: Cornell University Press, 1976.

Dubinsky, Ed. 1999. "Book Review: Mathematical Reasoning: Analogies, Metaphors, and Images." *Notices of the American Mathematical Society* 46, no. 5: 555–59.

Dzierzawa, Michael, and Marie-José Oméro. 2000. "Statistics of Stable Marriages." *Physica A: Statistical Mechanics and Its Applications* 287, nos. 1–2: 321–33. doi:10.1016/S0378-4371(00)00344-7.

Edgar, Scott. 2014. "Hermann Cohen's Principle of the Infinitesimal Method: A Rationalist Interpretation." Online draft. https://www.academia.edu/6787903/Hermann_Cohens_Principle_of_the_Infinitesimal_Method_a_Rationalist_Interpretation.

Ehrhardt, Caroline. 2010. "A Social History of the 'Galois Affair' at the Paris Academy of Sciences (1831)." *Science in Context* 23, no. 1: 91–119. doi:10.1017/S0269889709990251.

———. 2011. "A Quarrel between Joseph Liouville and Guillaume Libri at the French Academy of Sciences in the Middle of the Nineteenth Century." *Historia Mathematica* 38, no. 3: 389–414. doi:10.1016/j.hm.2011.02.002.

Ehrlich, Philip. 2006. "The Rise of Non-Archimedean Mathematics and the Roots of a Misconception I: The Emergence of Non-Archimedean Systems of Magnitudes." *Archive for History of Exact Sciences* 60, no. 1. doi:10.1007/s00407-005-0102-4.

Einaudi, Luigi. 2006. "The Theory of Imaginary Money from Charlemagne to the French Revolution." In *Selected Economic Essays*, edited by Luca Einaudi, Riccardo Faucci, and Roberto Marchionatti, 153–81. Basingstoke, UK: Palgrave Macmillan.

Eisen, M. 1969. *Elementary Combinatorial Analysis*. New York: Gordon & Breach.

Ernest, Paul. 1998. *Social Constructivism as a Philosophy of Mathematics*. Albany, NY: SUNY Press.

———. 2006. "A Semiotic Perspective of Mathematical Activity: The Case of Number." *Educational Studies in Mathematics* 61, nos. 1–2: 67–101. doi:10.1007/s10649-006-6423-7.

———. 2010. "Mathematics and Metaphor." *Complicity* 7, no. 1: 98–104.

Euler, Leonhard. 2000. *Foundations of Differential Calculus.* New York: Springer.

Ewald, William, ed. 1996. *From Kant to Hilbert: A Source Book in the Foundations of Mathematics.* Oxford: Clarendon Press.

Ferraro, Giovanni. 2007. "Convergence and Formal Manipulation in the Theory of Series from 1730 to 1815." *Historia Mathematica* 34, no. 1: 62–88. doi:10.1016/j.hm. 2005.08.004.

Field, Hartry H. 1980. *Science without Numbers: A Defence of Nominalism.* Princeton, NJ: Princeton University Press.

Field, Judith Veronica. 2005. *Piero Della Francesca: A Mathematician's Art.* New Haven, CT: Yale University Press.

Fisch, Menachem. Forthcoming. *Creatively Undecided: Toward a History and Philosophy of Scientific Agency.* Chicago: Chicago University Press.

Fontenelle, Bernard de. 1727. *Eléments de la Géométrie de l'Infini. Suite des Mémoires de l'Académie Royale des Sciences.* Paris: L'Imprimerie Royale. http://gallica.bnf.fr /ark:/12148/bpt6k64762n.

Foucault, Michel. 2002. *Archaeology of Knowledge.* 2nd ed. New York: Routledge.

Franci, Raffaella, and Laura Toti Rigatelli. 1985. "Towards a History of Algebra from Leonardo of Pisa to Luca Pacioli." *Janus* 72: 17–82.

Freeman, Walter J. 2000. *How Brains Make Up Their Minds.* New York: Columbia University Press, 2000.

———. 2008. "Nonlinear Brain Dynamics and Intention According to Aquinas." *Mind and Matter* 6, no. 2: 207–34.

———. 2009. "The Neurobiological Infrastructure of Natural Computing: Intentionality." *New Mathematics and Natural Computation* 5, no. 1 (March 1): 19–29. doi:10.1142/ S1793005709001179.

Frege, Gottlob. 1984. *Collected Papers on Mathematics, Logic and Philosophy.* Oxford: B. Blackwell.

Fresán, Javier. 2009. "The Castle of Groups: Interview with Pierre Cartier." *European Mathematical Society Newsletter* 74, no. 4: 31–34.

Freudenthal, Gideon. 2013. *Salomon Maimon: Rational Dogmatist, Empirical Skeptic: Critical Assessments.* Dodrecht: Springer.

Fried, Michael N., and Sabetai Unguru. 2001. *Apollonius of Perga's Conica: Text, Context, Subtext.* Leiden: Brill.

Friend, Michèle. 2007. *Introducing Philosophy of Mathematics.* Stocksfield, UK: Acumen.

———. 2014. *Pluralism in Mathematics: A New Position in Philosophy of Mathematics.* Dordrecht: Springer.

Fries, Jakob F. 1822. *Die Mathematische Naturphilosophie.* Heidelberg: Mohr und Winter. https://books.google.co.il/books?id=SZlEAAAAcAAJ.

Gale, D., and L. S. Shapley. 1962. "College Admissions and the Stability of Marriage." *American Mathematical Monthly* 69, no. 1: 9–15. doi:10.2307/2312726.

Gare, Arran. 2011. "From Kant to Schelling and Process Metaphysics: On the Way to Ecological Civilization." *Cosmos and History: The Journal of Natural and Social Philosophy* 7, no. 2: 26–69.

George, Alexander, and Daniel J. Velleman. 2002. *Philosophies of Mathematics.* Malden, MA: Blackwell.

Gold, Bonnie. 2001. "Book Review: Where Mathematics Comes From: How the Embodied Mind Brings Mathematics into Being." *Mathematical Association of America Reviews.* http://www.maa.org/press/maa-reviews/where-mathematics-comes-from-how-the-embodied-mind-brings-mathematics-into-being.

Goldin, Gerald A. 2001. "Counting on the Metaphorical." *Nature* 413, no. 6851: 18–19. doi:10.1038/35092607.

Grosholz, Emily R. 2007. *Representation and Productive Ambiguity in Mathematics and the Sciences.* Oxford: Oxford University Press, 2007.

——. 2014. "Fermat's Last Theorem and the Logicians." In *From a Heuristic Point of View: Essays in Honor of Carlo Cellucci,* edited by E. Ippoliti, 147–62. Cambridge: Cambridge Scholars Publishing.

——. Forthcoming. *Starry Reckoning: Reference and Analysis in Mathematics and Cosmology.* New York: Springer.

Gusfield, Dan, and Robert W. Irving. 1989. *The Stable Marriage Problem: Structure and Algorithms.* Cambridge, MA: MIT Press.

Hacking, Ian. 2014. *Why Is There Philosophy of Mathematics at All?* Cambridge: Cambridge University Press.

Hadamard, Jacques. 1945. *The Mathematician's Mind: The Psychology of Invention in the Mathematical Field.* Princeton, NJ: Princeton University Press.

Hall, Philip. 2009. "On Representatives of Subsets." *Journal of the London Mathematical Society, Series 1,* 10, no. 1: 26–30. doi:10.1112/jlms/s1-10.37.26.

Haug, Espen Gaarder, and Nassim Nicholas Taleb. 2011. "Option Traders Use (Very) Sophisticated Heuristics, Never the Black-Scholes-Merton Formula." *Journal of Economic Behavior & Organization* 77, no. 2: 97–106. doi:10.1016/j.jebo.2010.09.013.

Heeffer, Albrecht. 2008. "On the Nature and Origin of Algebraic Symbolism." In *New Perspectives on Mathematical Practices: Essays in Philosophy and History of Mathematics,* edited by Bart Van Kerkhove, 1–27. Singapore: World Scientific.

——. 2011. "On the Curious Historical Coincidence of Algebra and Double-Entry Bookkeeping." In *Foundations of the Formal Sciences VII: Bringing Together Philosophy and Sociology of Science,* edited by Karen François, Benedikt Löwe, Thomas Müller, and Bart Van Kerkhove, 109–30. London: College Publications.

Hellman, Geoffrey. 1998. "Maoist Mathematics?" *Philosophia Mathematica* 6, no. 3: 334–45. doi:10.1093/philmat/6.3.334.

Henderson, David. 2002. "Book Review: Where Mathematics Comes From." *Mathematical Intelligencer* 24, no. 1: 75–78.

Hersh, Reuben. 1997. *What Is Mathematics, Really?* Oxford: Oxford University Press.

Høyrup, Jens. 2007. *Jacopo Da Firenze's Tractatus Algorismi and Early Italian Abbacus Culture.* Basel: Birkhäuser. http://public.eblib.com/EBLPublic/PublicView.do?ptiID=337900.

——. 2008. "What Did the Abbacus Teachers Really Do When They (Sometimes) Ended Up Doing Mathematics? An Investigation of the Incentives and Norms of a Distinct Mathematical Practice." In *New Perspectives on Mathematical Practices: Essays in Philosophy and History of Mathematics,* edited by Bart Van Kerkhove, 47–75. Singapore: World Scientific.

——. 2010. "Hesitating Progress—the Slow Development toward Algebraic Symbolization in Abbacus and Related Manuscripts, c. 1300 to c. 1550." In *Philosophical*

Aspects of Symbolic Reasoning in Early Modern Mathematics, edited by Albrecht Heeffer and Maarten Van Dyck, 3–56. London: College Publications.

———. 2012. *Sanskrit-Prakrit Interaction in Elementary Mathematics as Reflected in Arabic and Italian Formulations of the Rule of Three—and Something More on the Rule Elsewhere*. Berlin: Max-Planck-Institut für Wissenschaftsgeschichte. http://www.mpiwg-berlin.mpg.de/Preprints/P435.pdf.

Hughes, Barnabas. 1987. "An Early 15th-Century Algebra Codex: A Description." *Historia Mathematica* 14, no. 2: 167–72. doi:10.1016/0315-0860(87)90020-6.

Immorlica, Nicole, and Mohammad Mahdian. 2005. "Marriage, Honesty, and Stability." In *Proceedings of the Sixteenth Annual ACM-SIAM Symposium on Discrete Algorithms*, 53–62. SODA '05. Philadelphia, PA: Society for Industrial and Applied Mathematics.

"Is the Black-Scholes Formula Just Plain Wrong?" 2014. *Quora*. http://www.quora.com/Is-the-Black-Scholes-formula-just-plain-wrong.

Jahnke, Hans Niels. 2003. "Algebraic Analysis in the 18th Century." In *A History of Analysis*, edited by Hans Niels Jahnke, 105–36. Providence, RI: American Mathematical Society.

Kato, Akiko. 1993. "Complexity of the Sex-Equal Stable Marriage Problem." *Japan Journal of Industrial and Applied Mathematics* 10, no. 1: 1–19. doi:10.1007/BF03167200.

Kaufman, Stav. 2016. "On the Emergence of a New Mathematical Object: An Ethnography of a Duality Transform." In *Mathematical Cultures*, edited by Brendan Larvor, 91–110. Basel: Birkhäuser.

Keller, Evelyn Fox. 1992. "Language and Ideology in Evolutionary Theory." In *Secrets of Life, Secrets of Death: Essays on Language, Gender, and Science*, 112–42. New York: Routledge.

Klein, Ursula. 2003. *Experiments, Models, Paper Tools: Cultures of Organic Chemistry in the Nineteenth Century*. Stanford, CA: Stanford University Press.

Knuth, Donald Ervin. 1997. *Stable Marriage and Its Relation to Other Combinatorial Problems: An Introduction to the Mathematical Analysis of Algorithms*. Providence, RI: American Mathematical Society.

Kövecses, Zoltán. 2002. *Metaphor: A Practical Introduction*. Oxford: Oxford University Press.

Krömer, Ralf, Colin McLarty, and Michael Wright. 2009. "Mini-Workshop: Category Theory and Related Fields: History and Prospects." *Oberwolfach Reports* 6, no. 1: 463–92. doi:10.4171/OWR/2009/08.

Lagrange, Joseph-Louis. 1797. *Théorie des Fonctions Analytiques*. Paris: L'Imprimerie de la République. http://gallica.bnf.fr/ark:/12148/bpt6k86263h.

Lakatos, Imre. 1976. *Proofs and Refutations: The Logic of Mathematical Discovery*. Cambridge: Cambridge University Press.

Lakoff, George. 2008. "The Neural Theory of Metaphor." In *The Cambridge Handbook of Metaphor and Thought*, edited by Raymond W. Gibbs. Cambridge Handbooks in Psychology. Cambridge: Cambridge University Press. doi:10.1017/CBO9780511816802.003.

Lakoff, George, and Rafael E. Núñez. 2000. *Where Mathematics Comes From: How the Embodied Mind Brings Mathematics into Being*. New York: Basic Books.

La Nave, Federica, and Barry Mazur. 2002. "Reading Bombelli." *Mathematical Intelligencer* 24, no. 1: 12–21. doi:10.1007/BF03025306.

Lando, Sergei K. 2003. *Lectures on Generating Functions*. Providence, RI: American Mathematical Society.

Landy, David, Colin Allen, and Carlos Zednik. 2014. "A Perceptual Account of Symbolic Reasoning." *Frontiers in Psychology* 5: 275. doi:10.3389/fpsyg.2014.00275.

Latour, Bruno. 2008. "Review Essay: The Netz-Works of Greek Deductions." *Social Studies of Science* 38, no. 3: 441–59. doi:10.1177/0306312707087973.

———. 2013. *An Inquiry into Modes of Existence: An Anthropology of the Moderns*. Cambridge, MA: Harvard University Press.

Lefebvre, Muriel. 2002. "Construction et Déconstruction des Diagrammes de Dynkin." *Actes de la recherche en sciences sociales* 141–42, no. 1: 121–26. doi:10.3917/arss.141.0121.

Lévi-Strauss, Claude. 1987. *Introduction to the Work of Marcel Mauss*. London: Routledge & Kegan Paul.

Liu, Chung L. 1968. *Introduction to Combinatorial Mathematics*. New York: McGraw-Hill College.

Livingston, Eric. 1986. *The Ethnomethodological Foundations of Mathematics*. London: Routledge & Kegan Paul, 1986.

———. 1999. "Cultures of Proving." *Social Studies of Science* 29, no. 6: 867–88. doi:10.1177/030631299029006003.

Ma, Jinpeng. 1996. "On Randomized Matching Mechanisms." *Economic Theory* 8, no. 2: 377–81. doi:10.1007/BF01211824.

MacBride, Elna B. 1971. *Obtaining Generating Functions*. New York: Springer.

MacMahon, Percy Alexander. 1915. *Combinatory Analysis*. Cambridge: Cambridge University Press. http://archive.org/details/combinatoryanaly02macmuoft.

Madden, James J. 2001. "Book Review: Where Mathematics Comes From: How the Embodied Mind Brings Mathematics into Being." *Notices of the American Mathematical Society* 48, no. 10: 1182–88. doi:10.2307/3072449.

Maddy, Penelope. 2007. *Second Philosophy: A Naturalistic Method*. Oxford: Clarendon Press.

———. 2011. *Defending the Axioms: On the Philosophical Foundations of Set Theory*. Oxford: Oxford University Press.

Mancosu, Paolo. 1999. *Philosophy of Mathematics and Mathematical Practice in the Seventeenth Century*. Oxford: Oxford University Press.

Mancosu, Paolo, ed. 2011. *The Philosophy of Mathematical Practice*. Oxford: Oxford University Press.

Martin, Emily. 1991. "The Egg and the Sperm: How Science Has Constructed a Romance Based on Stereotypical Male-Female Roles." *Signs* 16, no. 3: 485–501.

Martínez, Alberto A. 2006. *Negative Math: How Mathematical Rules Can Be Positively Bent*. Princeton, NJ: Princeton University Press.

Mason, John. 2010. "Your Metaphor, or My Metonymy?" *Complicity* 7, no. 1: 32–38.

McVitie, D. G., and L. B. Wilson. 1971. "The Stable Marriage Problem." *Communications of the Association for Computing Machinery* 14, no. 7: 486–90. doi:10.1145/362619.362631.

Merton, Robert C. 1973. "Theory of Rational Option Pricing." *Bell Journal of Economics and Management Science* 4, no. 1 (April 1): 141–83. doi:10.2307/3003143.

Mezard, Marc, Giorgio Parisi, and Miguel A Virasoro. 1997. *Spin Glass Theory and Beyond: An Introduction to the Replica Method and Its Applications*. Singapore: World Scientific.

Mill, John Stuart. 1843. *A System of Logic, Ratiocinative and Inductive*. London: J. W. Parker. https://books.google.co.il/books?id=y4MEAAAAQAAJ.

Mowat, Elizabeth, and Brent Davis. 2010. "Interpreting Embodied Mathematics Using Network Theory: Implications for Mathematics Education." *Complicity* 7, no. 1: 1–31.

Mueller, Ian. 1981. *Philosophy of Mathematics and Deductive Structure in Euclid's Elements*. Cambridge, MA: MIT Press.

Murawski, Roman. 2010. *Essays in the Philosophy and History of Logic and Mathematics*. Amsterdam: Rodopi.

Netz, Reviel. 1999. *The Shaping of Deduction in Greek Mathematics: A Study in Cognitive History*. Cambridge: Cambridge University Press.

———. 2004. *The Transformation of Mathematics in the Early Mediterranean World: From Problems to Equations*. Cambridge: Cambridge University Press.

Núñez, Rafael E. 2005. "Creating Mathematical Infinities: Metaphor, Blending, and the Beauty of Transfinite Cardinals." *Journal of Pragmatics* 37, no. 10: 1717–41. doi: 10.1016/j.pragma.2004.09.013.

Palmquist, Stephen. 1993. *Kant's System of Perspectives: An Architectonic Interpretation of the Critical Philosophy*. Lanham, MD: University Press of America.

Peirce, Charles Sanders. 1931–61. *Collected Papers of Charles Sanders Peirce*. Edited by Charles Hartshorne and Paul Weiss. Cambridge, MA: Harvard University Press.

Pfeifer, Rolf, and Josh Bongard. 2006. *How the Body Shapes the Way We Think: A New View of Intelligence*. Cambridge, MA: MIT Press.

Plato. 1973. *The Republic and Other Works*. Translated by Benjamin Jowett. New York: Anchor.

Poincaré, Henri. 1914. *Science and Method*. London: T. Nelson. http://archive.org/details/sciencemethod00poinuoft.

Potter, Elizabeth. 2001. *Gender and Boyle's Law of Gases*. Bloomington: Indiana University Press.

Presmeg, Norma C. 1992. "Prototypes, Metaphors, Metonymies and Imaginative Rationality in High School Mathematics." *Educational Studies in Mathematics* 23, no. 6: 595–610. doi:10.1007/BF00540062.

Putnam, Hilary. 2004. *Ethics without Ontology*. Cambridge, MA: Harvard University Press.

Quine, Willard Van Orman. 1948. "On What There Is." *Review of Metaphysics* 2, no. 1: 21–38.

Rashed, Roshdi. 1994. *The Development of Arabic Mathematics: Between Arithmetic and Algebra*. Dordrecht: Kluwer Academic.

Rav, Yehuda. 2008. "The Axiomatic Method in Theory and in Practice." *Logique et Analyse* 51, no. 202: 125–47.

Reck, Erich H., and Michael P. Price. 2000. "Structures and Structuralism in Contemporary Philosophy of Mathematics." *Synthese* 125, no. 3: 341–83. doi:10.1023/A:1005203923553.

Romero-Medina, Antonio. 2005. "Equitable Selection in Bilateral Matching Markets." *Theory and Decision* 58, no. 3: 305–24. doi:10.1007/s11238-005-6846-0.

Rosental, Claude. 2008. *Weaving Self-Evidence: A Sociology of Logic.* Princeton, NJ: Princeton University Press.

Roth, Alvin E., and Marilda A. O. Sotomayor. 1990. *Two-Sided Matching: A Study in Game-Theoretic Modeling and Analysis.* Cambridge: Cambridge University Press.

Roth, Alvin E., and John H. Vande Vate. 1990. "Random Paths to Stability in Two-Sided Matching." *Econometrica* 58, no. 6: 1475–80.

Rotman, Brian. 2000. *Mathematics as Sign: Writing, Imagining, Counting.* Stanford, CA: Stanford University Press.

Royal Swedish Academy of Sciences. 1997. "Nobel Prize in Economic Sciences Press Release." http://www.nobelprize.org/nobel_prizes/economic-sciences/laureates/1997/press.html.

Rudich, Steven. 2003. "The Mathematics of 1950s Dating: Who Wins the Battle of the Sexes?" Microsoft PowerPoint presentation, "Great Theoretical Ideas in Computer Science," Carnegie Mellon University, Pittsburgh, PA. http://www.cs.cmu.edu/afs/cs.cmu.edu/academic/class/15251/discretemath/Lectures/dating.ppt.

Russell, Bertrand. 1938. *The Principles of Mathematics.* 2nd ed. New York: W. W. Norton & Company. http://archive.org/details/principlesofmath005807mbp.

Sandifer, C. Edward. 2007. *How Euler Did It.* Washington, DC: Mathematical Association of America.

Schilpp, Paul Arthur, ed. 1949. *Albert Einstein, Philosopher-Scientist.* Evanston, IL: Library of Living Philosophers.

Schlimm, Dirk. "Conceptual Metaphors and Mathematical Practice: On Cognitive Studies of Historical Developments in Mathematics." *Topics in Cognitive Science* 5, no. 2: 283–98. doi:10.1111/tops.12018.

Schubring, Gert. 2005. *Conflicts between Generalization, Rigor, and Intuition.* New York: Springer.

Sergeyev, Yaroslav D. 2013. *Arithmetic of Infinity.* 2nd ed. ePubMATIC.

Serres, Michel. 2007. *The Parasite.* Minneapolis: University of Minnesota Press.

Sesiano, Jacques. 1985. "The Appearance of Negative Solutions in Mediaeval Mathematics." *Archive for History of Exact Sciences* 32, no. 2: 105–50. doi:10.1007/BF00329870.

Shapiro, Stewart, ed. 2005. *The Oxford Handbook of Philosophy of Mathematics and Logic.* Oxford: Oxford University Press.

Sigler, Laurence E. 2002. *Fibonacci's Liber Abaci.* New York: Springer.

Simons, Peter. 2001. "Review of M. Steiner: The Applicability of Mathematics as a Philosophical Problem." *British Journal for the Philosophy of Science* 52, no. 1: 181–84. doi:10.1093/bjps/52.1.181.

Sinclair, Nathalie, and Martin Schiralli. 2003. "A Constructive Response to 'Where Mathematics Comes From.'" *Educational Studies in Mathematics* 52, no. 1: 79–91.

Singmaster, David. 2004. *Source in Recreational Mathematics: An Annotated Bibliography.* 8th preliminary ed. http://puzzlemuseum.com/singma/singma6/SOURCES/singma-sources-edn8-2004-03-19.htm.

Srivastava, H. M., and H. L. Manocha. 1984. *A Treatise on Generating Functions.* Chichester, UK: E. Horwood.

Stedall, Jacqueline A. 2008. *Mathematics Emerging: A Sourcebook 1540–1900*. Oxford: Oxford University Press.

Steiner, Mark. 1998. *The Applicability of Mathematics as a Philosophical Problem*. Cambridge, MA: Harvard University Press.

———. 2001. "Review of a Subject with No Object: Strategies for Nominalistic Interpretation of Mathematics by John P. Burgess and Gideon Rosen." *Iyyun: The Jerusalem Philosophical Quarterly* 50: 73–76.

Swetz, Frank. 1987. *Capitalism and Arithmetic: The New Math of the 15th Century*. Translated by David Eugene Smith. La Salle, IL: Open Court Publishing.

Talagrand, Michel. 2003. *Spin Glasses: A Challenge for Mathematicians: Cavity and Mean Field Models*. New York: Springer.

Tanner, R.C.H. 1980a. "The Alien Realm of the Minus: Deviatory Mathematics in Cardano's Writings." *Annals of Science* 37, no. 2 (1980a): 159–78. doi:10.1080/00033798000200171.

———.1980b. "The Ordered Regiment of the Minus Sign: Off-Beat Mathematics in Harriot's Manuscripts." *Annals of Science* 37, no. 2: 127–58. doi:10.1080/00033798000200161.

Teo, Chung-Piaw, and Jay Sethuraman. 1998. "The Geometry of Fractional Stable Matchings and Its Applications." *Mathematics of Operations Research* 23, no. 4: 874–91. doi: 10.1287/moor.23.4.874.

Teo, Chung-Piaw, Jay Sethuraman, and Wee-Peng Tan. 1999. "Gale-Shapley Stable Marriage Problem Revisited: Strategic Issues and Applications (Extended Abstract)." In *Integer Programming and Combinatorial Optimization*, edited by Gérard Cornuéjols, Rainer E. Burkard, and Gerhard J. Woeginger, 429–38. Berlin: Springer.

Thurston, William P. 1994. "On Proof and Progress in Mathematics." *Bulletin of the American Mathematical Society* 30, no. 2: 161–77. doi:10.1090/S0273-0979-1994-00502-6.

Tucker, Alan. 2002. *Applied Combinatorics*. New York: Wiley, 2002.

Tymoczko, Thomas. 1998. *New Directions in the Philosophy of Mathematics: An Anthology*. Princeton, NJ: Princeton University Press.

Ulivi, Elisabetta. 2002. "Scuole e Maestri d'Abaco in Italia tra Medievo e Rinascimento." In *Un Ponte sul Mediterraneo*, edited by Enrico Giusti and Raffaella Petti, 121–59. Firenze: Polistampa.

———. 2008."Scuole d'Abaco e Insegnamento della Matematica." In *Il Rinascimento Italiano e l'Europa*, edited by Antonio Clericuzio, Maria Conforti, and Germana Ernst, vol. 5, *Le Scienze*, 403–20. Costabissara: Angelo Colla.

Unguru, Sabetai, and David E. Rowe. 1981. "Does the Quadratic Equation Have Greek Roots? A Study of 'Geometric Algebra,' 'Application of Areas,' and Related Problems (I/II)." *Libertas Mathematica* 1: 1–49. doi:10.14510/lm.v1i0.354.

———. 1982. "Does the Quadratic Equation Have Greek Roots? A Study of 'Geometric Algebra,' 'Application of Areas,' and Related Problems (II/II)." *Libertas Mathematica* 2: 1–62. doi:10.14510/lm.v1i0.354.

Vande Vate, John H. 1989. "Linear Programming Brings Marital Bliss." *Operations Research Letters* 8, no. 3: 147–53. doi:10.1016/0167-6377(89)90041-2.

Van Egmond, Warren. 1980. *Practical Mathematics in the Italian Renaissance: A Catalog of Italian Abbacus Manuscripts and Printed Books to 1600*. Firenze: Parenti.

———. 1983. "The Algebra of Master Dardi of Pisa." *Historia Mathematica* 10, no. 4 (1983): 399–421. doi:10.1016/0315-0860(83)90003-4.

Van Kerkhove, Bart, ed. 2009. *New Perspectives on Mathematical Practices: Essays in Philosophy and History of Mathematics.* Singapore: World Scientific.

Van Kerkhove, Bart, and Jean Paul Van Bendegem, eds. 2007. *Perspectives on Mathematical Practices: Bringing Together Philosophy of Mathematics, Sociology of Mathematics, and Mathematics Education.* Dordrecht: Springer.

Wagner, Roy. 2009a. "For Some Histories of Greek Mathematics." *Science in Context* 22, no. 4: 535–65. doi:10.1017/S0269889709990159.

———. 2009b. "Mathematical Marriages: Intercourse between Mathematics and Semiotic Choice." *Social Studies of Science* 39, no. 2: 289–308. doi:10.1177/0306312708099443.

———. 2009c. "Mathematical Variables as Indigenous Concepts." *International Studies in the Philosophy of Science* 23, no. 1: 1–18. doi:10.1080/02698590902843351.

———. 2010a. "For a Thicker Semiotic Description of Mathematical Practice and Structure." In *Philosophy of Mathematics: Sociological Aspects and Mathematical Practice*, edited by Benedikt Löwe and Thomas Müller, 361–84. London: College Publications.

———. 2010b. "The Geometry of the Unknown: Bombelli's Algebra Linearia." In *Philosophical Aspects of Symbolic Reasoning in Early Modern Mathematics*, edited by Albrecht Heeffer and Maarten Van Dyck, 229–69. London: College Publications.

———. 2010c. "The Natures of Numbers in and around Bombelli's *L'algebra.*" *Archive for History of Exact Sciences* 64, no. 5: 485–523. doi:10.1007/s00407-010-0062-1.

———. 2012. "Infinity Metaphors, Idealism, and the Applicability of Mathematics." *Iyyun: The Jerusalem Philosophical Quarterly* 61: 129–48.

———. 2013. "A Historically and Philosophically Informed Approach to Mathematical Metaphors." *International Studies in the Philosophy of Science* 27, no. 2: 109–35. doi:10.1080/02698595.2013.813257.

———. 2014. "Wronski's Infinities." *History of Philosophy of Science* 4: 26–61.

———. Forthcoming. "Wronski's Foundations." *Science in Context.*

Walkerdine, Valerie. 1977. "Redefining the Subject in Situated Cognition Theory." In *Situated Cognition: Social, Semiotic, and Psychological Perspectives*, edited by David Kirshner and James A. Whitson, 57–70. Mahwah, NJ: Lawrence Erlbaum Associates.

Walsh, Vincent. 2003. "A Theory of Magnitude: Common Cortical Metrics of Time, Space and Quantity." *Trends in Cognitive Sciences* 7, no. 11 (November): 483–88. doi:10.1016/j.tics.2003.09.002.

West, Douglas Brent. 1996. *Introduction to Graph Theory.* Upper Saddle River, NJ: Prentice Hall.

Weyl, Hermann. 1949. "Almost Periodic Invariant Vector Sets in a Metric Vector Space." *American Journal of Mathematics* 71, no. 1 (January 1): 178–205. doi:10.2307/2372104.

Wigner, Eugene P. 1960. "The Unreasonable Effectiveness of Mathematics in the Natural Sciences." *Communications on Pure and Applied Mathematics* 13, no. 1: 1–14. doi:10.1002/cpa.3160130102.

Wilf, Herbert S. 1994. *Generatingfunctionology.* 2nd ed. San Diego, CA: Academic Press.

Wilson, Margaret. 2002. "Six Views of Embodied Cognition." *Psychonomic Bulletin & Review* 9, no. 4: 625–36. doi:10.3758/BF03196322.

Wilson, Robin J. 1996. *Introduction to Graph Theory*. 4th ed. Harlow: Longman, 1996.

Wittgenstein, Ludwig. 2001. *Philosophical Investigations: The German Text, with a Revised English Translation*. Translated by G. E. M. Anscombe. 3rd ed. Malden, MA: Blackwell.

———. 1976. *Wittgenstein's Lectures on the Foundations of Mathematics, Cambridge, 1939*. Chicago: University of Chicago Press.

Wood, David W. 2012. *Mathesis of the Mind: A Study of Fichte's Wissenschaftslehreand Geometry*. Amsterdam: Rodopi.

Wöpcke, Franz. 1851. *L'algèbre d'Omar Alkhayyâmî*. Paris: B. Duprat. https://archive .org/details/lalgbredomaralk00khaygoog.

Index

■ ■ ■ ■ ■